AF587889

ETHICAL ISSUES IN THE 21ST CENTURY

BIOETHICS: ISSUES AND DILEMMAS

ETHICAL ISSUES IN THE 21ST CENTURY

Additional books in this series can be found on Nova's website under the Series tab.

Additional E-books in this series can be found on Nova's website under the E-books tab.

ETHICAL ISSUES IN THE 21ST CENTURY

BIOETHICS: ISSUES AND DILEMMAS

TYLER N. PACE
EDITOR

Nova Science Publishers, Inc.
New York

For permission to use material from this book please contact us:
Telephone 631-231-7269; Fax 631-231-8175
Web Site: http://www.novapublishers.com

LIBRARY OF CONGRESS CATALOGING-IN-PUBLICATION DATA

Bioethics : issues and dilemmas / editor, Tyler N. Pace.
p. ; cm.
Includes bibliographical references and index.
ISBN 978-1-61728-290-4 (hardcover)
1. Medical ethics. 2. Bioethics. I. Pace, Tyler N.
[DNLM: 1. Bioethical Issues. 2. Ethics, Medical. WB 60 B615473 2010]
R724.B4523 2010
174.2--dc22

2010022564

Published by Nova Science Publishers, Inc. † New York

Contents

PREFACE

Bioethics is the philosophical study of the ethical controversies brought about by advances in biology and medicine. Bioethicists are concerned with the ethical questions that arise in the relationships among life sciences, biotechnology, medicine, politics, law, philosophy, and theology. This new book presents research in the expansive field of bioethics including biomedical ethics in obstetrics, ethical decision making in the health care system, the feasibility of using human oocytes for stem cell research, as well as mandatory circumcision in Sub-Saharan Africa to prevent HIV and AIDS and environmental ethics to preserve the world for future generations.

Chapter 1 - Human Immunodeficiency Virus (HIV) and Acquired Immunodeficiency Syndrome (AIDS) are pandemic and are causing unprecedented devastation on societies and economies throughout the world. There is no place in the world that has felt this devastation more than sub-Saharan Africa. The United Nations Programme on HIV/AIDS (UNAIDS) and the World Health Organization (WHO) estimate that at the end of 2007 there were an estimated 33 million [30 million to 36 million] people living with HIV. The annual number of new HIV infections declined from 3.0 million [2.6 million to 3.5 million] in 2001 to 2.7 million [2.2 million to 3.2 million] in 2007. Approximately 67% of people living with HIV and 72% of AIDS deaths in 2007 came from sub-Saharan Africa. Globally, the number of children 15 years of age and younger living with HIV increased from 1.6 million [1.4 million to 2.1 million] in 2001 to 2.0 million [1.9 million to 2.3 million] in 2007. Almost 90% live in sub-Saharan Africa. Two-thirds (63%) of all adults and children with HIV globally live in sub-Saharan Africa. [1] Recent studies have shown that in some sub-Saharan countries the HIV infection trends appear to be stable. Experts explain that “in essence this reflects equilibrium:

the number of people newly infected with HIV roughly equal the number of people dying of AIDS." [2] Medically and ethically, this is unacceptable. Lives are being lost daily when there are possible procedures available that might lessen the number of infections and deaths.

Chapter 2 - Medical ethics consist fundamentally in using rational knowledge to ensure the respect for human dignity. Medicine has always been in its very essence an ethical process as evinced by the ancient Hippocratic oath. Nowadays, ethical issues must be placed within the context of the changes in our way of doing things due to the extraordinary evolution of science and technology. In the field related to the beginning of life, progress in prenatal diagnosis has completely changed our medical approach to pregnancy. This has undeniably become one of the recurring themes in bioethical debates in France, alongside questions about medically assisted procreation, surrogacy and stem cell research. Indeed, France has generally been considered a pioneer in this field since its first bioethical laws were adopted in 1994. Having reviewed these laws three times in fifteen years, the French legislature has always deemed it necessary to supervise biomedical practices so as to avoid reaching a point of no return. However, with medical ethics defending the very principle of humanity on the one hand and a bioethical arsenal of legislation on the other, may there not be a risk of losing coherence? Based on the daily practice of an obstetrician-gynecologist, the authors of this chapter raise questions about the possible conflicts of values, and possible irrational aspects they might reveal.

Chapter 3 - In recent years the literature on bioethics has begun to pose the sociological challenge of how to explore organisational processes that facilitate a systemic response to ethical concerns. In this chapter, I seek to make a contribution to this important new direction in ethical research by presenting an overview of my program of research that explores notions that inform health professionals' ethical decision-making in the 'real world' of health care organisations. The insights gained through the research are translated into practical strategies that can be utilized to strengthen the process of ethical decision-making in the health care system.

Chapter 4 - Today it is widely recognised that much of the harm done to the environment is not limited by geographical, political or time barriers.

From the point of view of environmental ethics, our concern for future generations arises mainly from their position of disadvantage: preceding generations can limit the opportunities bequeathed to later generations by causing irreversible harm and depleting resources.

This paper first introduces some historical references; the most significant problems posed by the rights of future generations in regard to the environment are then identified, the notions of "common heritage" and "capital" as referred to the environment are discussed and the relations between "sustainable development" and the rights of future generations are presented. Some social aspects, legal implications and the contributions of philosophers and experts on the question of intergenerational justice in environmental terms are then analysed. An attempt is finally made to define the most significant problems involved.

Chapter 5 - Most research within biomedical ethics consists of theoretical reflections regarding which ethical theories or principles are useful to analyse ethical issues in the field of biomedicine. The theories of the American ethicists Tom L. Beauchamp & James F. Childress (2009) and the Danish philosophers Jakob Rendtorff & Peter Kemp (2000) are examples. Beauchamp & Childress examined considered moral judgements and the way moral beliefs cohere and found that the general principles of beneficence, nonmaleficence, respect for autonomy, and justice play a vital role in biomedical ethics (Beauchamp & Childress, 2009, p. 13). These authors believe that the four principles are not only specific for biomedical ethics, they are found in all cultures in everyday life because they are part of a cross-cultural common morality shared by all persons committed to morality (Beauchamp & Childress, 2009, p. 4). The four clusters of principles provide a framework of norms to start with in biomedical ethics. As a starting point no principle is weighted higher than the other principles. When occasion arises, the principles are weighted, balanced, and specified. As will become clear, Rendtorff & Kemp do not find the approach of Beauchamp & Childress convincing, and they developed an alternative theory based on the following four basic ethical principles: autonomy, dignity, integrity, and vulnerability. Rendtorff & Kemp regard Beauchamp & Childress' approach as individualistic, and they believe that their alternative fills out an empty space in the American theory by protecting "the fragile and finite, bodily incarnated human person" and thereby leads to a wider view of the human person (Rendtorff & Kemp, 2000, p. 314).

First, this article presents a philosophical analysis of the bioethical theory of Beauchamp & Childress. Next, the theory of Beauchamp & Childress is discussed and compared it with the theory of Rendtorff & Kemp.

Chapter 6 - Human embryonic stem cell research aimed at producing patient-specific tissues for cell-based therapies needs to create embryos through somatic cell nuclear transfer (SCNT), for which human oocytes will be necessary. However, oocyte donation is a risky and uncomfortable process, partly because of hormonal stimulation to induce ovulation. *In vitro* maturation (IVM) of immature human oocytes is in clinical use for infertility treatments. This article discusses the feasibility of, and ethical and legal considerations for, using IVM of human oocytes retrieved from surgically removed ovaries for SCNT; and concludes IVM of human oocytes from surgically removed ovaries clearly warrants exploration.

In: Bioethics: Issues and Dilemmas
Editor: Tyler N. Pace
ISBN: 978-1-61728-290-4

Chapter 1

SURGICAL VACCINE: SHOULD MALE CIRCUMCISION BE MANDATORY IN SUB-SAHARAN AFRICA

Peter A. Clark[1], Justin Eisenman[2] and Stephen Szapor[1]
[1]Saint Joseph's University, Philadelphia, Pennsylvania
[2]Madigan Army Medical Center, Fort Lewis, Washington

INTRODUCTION

Human Immunodeficiency Virus (HIV) and Acquired Immunodeficiency Syndrome (AIDS) are pandemic and are causing unprecedented devastation on societies and economies throughout the world. There is no place in the world that has felt this devastation more than sub-Saharan Africa. The United Nations Programme on HIV/AIDS (UNAIDS) and the World Health Organization (WHO) estimate that at the end of 2007 there were an estimated 33 million [30 million to 36 million] people living with HIV. The annual number of new HIV infections declined from 3.0 million [2.6 million to 3.5 million] in 2001 to 2.7 million [2.2 million to 3.2 million] in 2007. Approximately 67% of people living with HIV and 72% of AIDS deaths in 2007 came from sub-Saharan Africa. Globally, the number of children 15 years of age and younger living with HIV increased from 1.6 million [1.4 million to 2.1 million] in 2001 to 2.0 million [1.9 million to 2.3 million] in

2007. Almost 90% live in sub-Saharan Africa. Two-thirds (63%) of all adults and children with HIV globally live in sub-Saharan Africa. [1] Recent studies have shown that in some sub-Saharan countries the HIV infection trends appear to be stable. Experts explain that "in essence this reflects equilibrium: the number of people newly infected with HIV roughly equal the number of people dying of AIDS." [2] Medically and ethically, this is unacceptable. Lives are being lost daily when there are possible procedures available that might lessen the number of infections and deaths.

The majority of those infected with HIV in sub-Saharan Africa do not have access to antiretroviral therapy, which is known to prolong the lives of HIV-positive persons in industrialized countries. Although the availability of antiretroviral therapy for those infected with HIV has increased worldwide, the infection rate out surpasses those started on such treatment. 3]Provision of antiretroviral therapy has expanded dramatically in sub-Saharan Africa; however, the numbers needing such therapy are staggering. The estimated number of people receiving antiretroviral therapy is 2.1 million [1.92 to 2.31]; the estimated number of people needing antiretroviral therapy is 7 million [6.25 to 7.90]; antiretroviral therapy coverage is 30% of sub-Saharan Africa. [4] Without an AIDS vaccine or curative treatment, and given the difficulty in getting persons at risk to adopt healthy sexual behaviors, alternative approaches to decrease the spread of HIV infection are urgently needed. Multiple observational studies over the years in sub-Saharan Africa correlated male circumcision with a reduced risk of HIV infection. [5] Three recent randomized controlled trials undertaken in Kisumu, Kenya, [6] Rakai District, Uganda [7]and Orange Farm, South Africa [8] have confirmed that male circumcision reduces the risk of heterosexually acquired HIV infection in men by approximately 51% to 60%. These three studies provide a solid evidence-base for future health policy. In March 2007 both WHO and UNAIDS recommended circumcision to reduce male heterosexual HIV acquisition. Both organizations estimate that male circumcision could prevent 5.7 million sub-Saharan African men from contracting HIV over the next two decades and save possibly 3 million lives. Currently, an estimated 665 million men, or 30% of men worldwide, are estimated to be circumcised. [9]

Biological studies support a plausible mechanism for this protection. Compared to the dry external skin surface, the inner mucosa of the foreskin has less keratinization (deposition of fibrous protein), a higher density of target cells for HIV infection (Langerhans cells), and is more susceptible to HIV infection in laboratory studies. Studies have also shown that the foreskin may have greater susceptibility to traumatic epithelial disruptions (tears) during intercourse, providing a portal of entry for pathogens including HIV. In addition, the micro-environment in the preputial sac between the unretracted foreskin and the glans penis may be conducive to viral survival. Finally, the higher rates of sexually transmitted genital ulcerative disease, such as syphilis, observed in uncircumcised men may also increase susceptibility to HIV infection. [10] Over the years, researchers have found that HIV tends to be less prevalent in areas where male circumcision is common, however; it must be emphasized that male circumcision whether done neonatally or as an adolescent or adult is not a cure-all, because it only lessens the chances that a male will be infected by HIV. Neonatally, the procedure is relatively inexpensive and the risks diminish considerably. The procedure for adolescents and adults is expensive compared to abstinence, condoms or other methods; and the surgery is not without serious risks if performed by traditional healers using unsterilized blades as often happens in rural Africa. However, if the procedure is done by trained medical professionals the risks diminish considerably. Despite these risks, the growing scientific evidence shows that male circumcision reduces the risk of males contracting HIV and that circumcised men are about 30% less likely to transmit HIV to their female partners. Weighing the benefits and burdens, should male circumcision, also being called the "surgical vaccine," become mandatory in sub-Saharan Africa or should it be made universally available and aggressively promoted? However, before making this decision, various ethical concerns as well as complex traditional, cultural and religious concerns will have to be addressed in order for countries to incorporate this procedure into a national health care policy.

This article will focus on sub-Saharan Africa because it is the single area of the world that is being most devastated by the HIV/AIDS pandemic. However, the arguments can easily be applied to other areas of the world as well. Having spent time working in a medical clinic in Dar es Salaam, Tanzania with three medical students in the summer of 2007, it became clear that more has to be done to help decrease the spread of HIV in sub-Saharan Africa. There must be a comprehensive plan that encompasses not only testing, education and antiretroviral therapy but also new initiatives such as male circumcision. The intended purpose of this article is threefold: first, to examine the medical evidence regarding the effect of mandatory male circumcision on the spread of HIV/AIDS; second, to give an ethical analysis of the arguments for and against mandatory male circumcision; and third, to develop guidelines to implement male circumcision in sub-Saharan Africa as a strategic health policy.

Medical Analysis

Male circumcision is the removal of some or all of the foreskin of the penis. [11] An uncircumcised penis consists of a cylindrical shaft and a rounded tip (glans) which is separated by a tissue groove called the coronal sulcus. During the circumcision procedure the foreskin is removed to a point near the coronal sulcus. [12]Male circumcision is believed to be one of the oldest forms of surgery and the earliest evidence for circumcision comes from Egypt where tomb artwork depicts uncircumcised men. [13] One writer notes that "the Egyptians, who probably acquired the practice from African tribes, practiced circumcision as early as 4000 B.C. In the Jewish religion, the origin of circumcision is attributed to Abraham, who established the 'blood covenant' with God through circumcision. This practice is also of pre-Islamic Arabic tradition and became a near prerequisite to becoming a Moslem. The practice is very widespread and is unknown only to Indo-Germanic people, the Mongols, and non-Moslem Finno-Ugrian races." [14] Historically, circumcision has been performed to prevent phimosis which is an inability to retract the foreskin. Medically the procedure has also been performed to protect a male from infant urinary tract infections, cancer of the penis, and other sexually transmitted diseases. [15] Traditionally there are other reasons for male circumcision. The procedure has been performed to enhance a man's sexual performance, sacrifice a piece of his flesh to fertility gods, demonstrate his ability to withstand pain, demonstrate his membership in a particular tribe, enable him to look like his circumcised father, and to gain social prestige. [16]

There are three different techniques of male circumcision which are performed on neonates: the Gomco clamp, the Plastibel device, and the Mogen clamp. In each of these techniques the surgeon estimates the amount of foreskin needed to be removed, dilates the preputial orfice exposing the glans to ensure it is normal, frees the inner preputial epithelium from the epithelium of the glans, places the device and leaves the device in situ long enough to produce hemostasis and amputation of the foreskin. [17] In the observational studies performed in Kisumu, Kenya and Rakai, Uganda, the sleeve procedure was used to circumcise the males. During this procedure the foreskin was retreated and an incision was made proximal to the coronal sulcus. A proximal incision was then made on the unretracted foreskin at the corona. Then the superficial lamina was exposed and the sleeve of the foreskin was freed and removed from the Bucks fascia. [18] In 2009, Clinical Innovations, a division of ACI Medical Devices, launched a new disposable circumcision device called AccuCirc. It provides physicians with a precise, consistent and reliable circumcision outcome. "The AccuCirc uses two innovative components to complete the circumcision procedure: the Foreskin Probe/Shielding Ring and the Single-Action Clamp. The Foreskin Probe/Shielding Ring ensures that the glans is protected and the foreskin is properly aligned, while the Single-Action Clamp ensures adequate hemostasis and the precise delivery of the protected, circular blade. These two components work together to protect the infant from injury." [19] The AccuCirc comes in a self-contained circumcision kit; including the foreskin probe/shielding ring, single-action clamp and blade, fenestrated drape with adhesive backing, surgical marking pen, hemostats, sanitizing wipe, iodine swab sticks, lubricating jelly, petrolatum dressing, and gauze. The kit is completely disposable.[20]

Adolescent and adult male circumcision is more complicated than infant circumcision, and is commonly performed using one of the following techniques: 1) guided forcepts, 2) dorsal slit, or 3) sleeve resection. Before surgery, the pubic area is thoroughly scrubbed, and pubic hair may be shaved or clipped. An anesthetic cream may be applied and a local anesthetic is injected into the base of the penis. This may be followed with additional injections in a ring around the shaft. The following is a brief description of the three techniques:

1) The guided forceps is the simplest technique. In this procedure, the foreskin is pulled forward over the glans with a pair of forceps, and the foreskin is then snipped, using the edge of the forceps as a guide.
2) The dorsal slit is often preferred when treating phimosis or paraphimosis. In this procedure, a slit is made from the opening of the foreskin to a point a few centimeters in, and then a circle is cut around the glans.
3) Sleeve resection is more complicated, but often preferred when there is a risk for excessive bleeding. In this procedure, two parallel cuts are made along the shaft of the penis, resulting in a thin band or sleeve of detached foreskin. When this is removed, the top and bottom portions of the foreskin are attached with dissolving sutures.[21]

Following circumcision, an antibiotic ointment is applied and the area is wrapped in loose gauze.

Globally, 70% of men infected with HIV contracted the virus through vaginal sex, and a small number acquired it from anal intercourse. [22] The biological mechanism behind circumcision and its effect on HIV rests in the cells found in the foreskin. The outer shaft of the penis is protected by a barrier of keratinized stratified squamous epithelium, much like the structure of skin found throughout the outer surfaces of the body. In these tissues, keratin provides a tough structural matrix that resist friction and fluids. The inner mucosa of the foreskin, however, is made up of non-keratinized squamous tissue and thus does not offer the same level of protection. During penetration, the foreskin is retracted down the shaft of the penis, exposing the inner mucosa to vaginal or anal secretions. [23] Furthermore, the inner mucosa is rich in antigen presenting cells. These cells include dendritic cells, macrophages and Langerhans' cells. These cells have cell-surface Fc (antibody) and C3 (complement) and act through phagocytosis of foreign antigens. In most cases of infection, HIV binds to the CD4 and CCR5 receptors found on antigen presenting cells, such as Langerhans' cells. [24] When these cells are infected with the virus, they adhere to adjacent CD4 lymphocytes and migrate into lymph nodes where they present epitopes of processed foreign antigens to T lymphocytes. The higher concentration of Langerhan cells of the inner foreskin of the uncircumcised penis, coupled with the lack of protective keratin, make the inner mucosa particularly vulnerable to inoculation by HIV. [25]

Aside from the high number of antigen presenting cells, the foreskin may also contribute to a higher rate of HIV infection due to the greater incidence of ulcerative sexually transmitted infections and the susceptibility of the foreskin to abrasion. HIV positive males often have pre-existing genital ulcers due to an untreated STD other than HIV. [26] It is believed that these lesions may provide an entry point for the HIV virus. This is compounded by the fact that in uncircumcised males, the frenulum, rich in blood supply, is particularly susceptible to injury during intercourse. In addition to this, the environment under the foreskin (moisture and temperature) may favor microorganism survival and replication. Thus, circumcision in males further reduces the synergistic effect seen between HIV and sexually transmitted infections. [27]

The impact of circumcision in males is supported by recent epidemiological evidence from multiple observational studies over the years in sub-Saharan Africa. Three recent randomized controlled trials undertaken in Kisumu, Kenya, Rakai District in Uganda and Orange Farm, South Africa have shown that male circumcision reduces the risk of heterosexually acquired HIV infection in men by approximately 51% to 60%. [28] The study conducted in Kisumu, Kenya by Robert Bailey and colleagues, randomly assigned participating men between the ages of 18 and 24 to a circumcision group (n=1391) or delayed circumcision group (n=1393). During an unscheduled interim analysis in December of 2006 it became apparent that circumcision produced a significant benefit and the study was halted. The two year HIV incidence rate was 2.1% in the circumcision group and 4.2% in the control group. There was an estimated 53% (unadjusted modified intention-to-treat analysis) to 60% (as-treated analysis) reduction in relative risk of HIV infection associated with male circumcision. [29]

In Rakai, Uganda, Ronald Gray and colleagues included 4996 men between the ages of 15 and 49. Of these participants, 2474 were randomized to the immediate circumcision group and 2522 were assigned to the delayed circumcision control group. Similar to the study in Kenya, after an interim analysis revealed the significant benefits, the Rakai study was stopped. HIV incidence was 0.66 per 100 person-years in the circumcision group and 1.33 per 100 person-years in the control group. The estimated reduction in the relative risk of infection with HIV was 51% (unadjusted modified intention-to-treat analysis) to 55% (as-treated analysis). Both trials were methodologically and analytically sound. [30]

In Orange Farm, South Africa, a suburb or Johannesburg, Bertran Auvert and colleagues enrolled 3274 men between the ages of 18 and 24. Of these participants, 1617 were randomized to immediate circumcision and 1657 to the delayed control group. After observing a 60% relative reduction in HIV risk during an interim analysis, the study was discontinued. Initially, concern was expressed about the randomization procedures in this study, the slight imbalance in baseline characteristics between groups, and potential selection bias, but in light of the Kenya and Uganda studies, the results from South Africa seem plausible. [31]

Over the next twenty years the WHO and UNAIDS estimate that male circumcision could prevent 5.7 million more infections, could prevent 300,000 deaths in the next ten years and could avert more than 3 million deaths over the next 20 years. [32] Male circumcision encompasses both neonatal and adolescent/adult procedures. The benefits and risks for both groups focus on issues of safety and cost. Neonatal male circumcision has some very important advantages when compared to circumcising as an adolescent or an adult. Mandating neonatal male circumcision could open up discussions about HIV prevention and would allow time for the children to be educated on subjects such as condom use, testing for HIV, mutual monogamy, and partner reduction as they get closer to the age of sexual activity. This education, along with the preventative effects of the circumcision, would cause a major decrease in the spread of the HIV virus. In addition, neonatal male circumcision is more effective if performed earlier in a man's life because the protective effect is greater due to the thickening of the skin on the head of the penis. [33] The procedure also minimizes the factor of risk compensation. This occurs when a person engages in a risky activity, such as having unprotected sex, due to a decrease in the perceived risk of the action, caused by the belief that male circumcision eliminates the possibility of acquiring HIV/AIDS. If males are circumcised at birth it is unlikely that male circumcision would have an effect on their risk perception of the preventative effects of circumcision, with regards to the HIV virus, when engaging in sexual activities. [34]

Neonatal male circumcision does minimize risk factors, but considering all the facts surrounding HIV in sub-Saharan Africa, public health officials are now arguing that circumcision for all men should be a key weapon in fighting HIV/AIDS in Africa. Numerous clinical studies have shown that male circumcision provides increased protection against human papillomavirus (HPV), herpes simplex virus, syphilis, and chancroid. In April 2009, clinicians reported that male circumcision reduces the acquisition of HPV and genital herpes (HSV-2) in men allocated to the intervention arm of a large randomized controlled trial. [35] In a similar study, Auvert et al. found that male circumcision reduces the prevalence of high-risk HPV and HIV in men. [36] Other benefits include reduced risks of genital cancer, urinary tract infections, elimination of phimosis, and paraphimosis and sexually transmitted infections in men and their female partners.[37] Complementing recent clinical trials in Africa, Warner et al. found that being circumcised significantly reduced the risk of HIV infection in heterosexual African American men known to be exposed to the virus. This new study suggests that male circumcision may protect heterosexual males in the United States as well. [38] One surprising benefit is that during the Orange Farm study in South Africa, the staff learned that nearly two-thirds of young women from the area reported that they preferred sex with circumcised men. This can only help the demand for the procedure. [39] It is clear that male circumcision has benefits throughout the course of one's life and does not decrease a man's pleasure or function. Neonatal male circumcision is safer, has a faster recovery rate, involves no stitches and costs less. Adolescent and adult circumcision allows for informed consent, but it does involve minimal risks, stitches, acute pain and is more expensive. Providing adequate education and training for health care professionals can reduce the risk of complications from male circumcision and can minimize concerns about the safety of this procedure. All things considered, the medical benefits of neonatal as well as adolescent/adult male circumcision certainly outweigh the minimal risks.

Cost factors also play a role in male circumcision in sub-Saharan Africa. Based on 2005 conditions in South Africa it was determined that male circumcision is highly cost-effective and would save approximately 2.4 million dollars over a twenty year period per 1000 circumcisions. [40] Because the procedure only needs to be performed once during a man's life, whether it is as an infant or as an adolescent/adult, it is a relatively inexpensive preventative technique. [41] "Major international funders, including the Bill and Melinda Gates Foundation and the U. S. President's Plan For AIDS Relief (PEPFAR), agree that ramped-up circumcision efforts must be funded as add-on services to guarantee that they will not detract from other programs." [42] Family Health International, a global public health and development organization, has won a five-year, $18.5 million grant from the Bill and Melinda Gates Foundation to boost male circumcision in Kenya. PEPFAR has granted $26 million for circumcision programs in 13 African countries—Botswana, Kenya, Rwanda, Zambia, South Africa, Lesotho, Malawi, Mozambique, Tanzania, Uganda, Namibia, Ethiopia, and Swaziland. The problem is that implementation has been highly variable. [43] For example, Swaziland is charging about $40 for a circumcision, but this is not an insignificant amount of money for many African men. [44] To overcome the cost factors international funding and government subsidization for the expense of male circumcision must become a reality. Male circumcision offers vaccine-level protection against HIV/AIDS especially in areas of high HIV prevalence, as is the case in sub-Saharan Africa. [45] Based on the fact that the procedure offers vaccine-level protection, one can argue convincingly that mandating male circumcision is not only medically necessary but is also economically justified similar to other vaccines that are mandated by many sub-Saharan African governments.

Non-Medical Issues Surrounding Mandatory Male Circumcision

There are also many ethnic, cultural and religious issues that need to be addressed when trying to determine if male circumcision should be made mandatory in sub-Saharan Africa. The first issue is determining whether or not male circumcision would be accepted among the people of sub-Saharan Africa. "Male circumcision rates in sub-Saharan Africa vary greatly; they tend to be highest in West Africa, where for example, an estimated 95% of men in Ghana and 93% of men in Cameroon are circumcised. In East and southern Africa, by comparison, an estimated 25% of men in Botswana and 35% in

South Africa are circumcised. There is also variation between ethnic groups within countries. In Kenya, for example, circumcision rates vary from 17% among Luo men to 100% among Somali men." [46] In one survey among non-circumcising communities 81% of couples said that they were willing to circumcise their children.[47] Some concerns were price, pain, and lack of complications. Pain can be minimized by local anesthesia and the cost of male circumcision could be integrated into existing medical costs or supported by international funders.[48] In Tanzania a male circumcision performed in a hospital costs around 10,000 shillings (approximately 8 US dollars). During the South African observational studies, circumcisions were performed for 300 South African Rand (approximately 40 dollars). [49] A cost-benefit analysis of male circumcision versus treatment for HIV/AIDS makes the male circumcision cost-effective.

Another issue confronting male circumcision focuses on religious and cultural practices. Currently the practice of male circumcision in Sub-Saharan Africa differs among religious and ethnic groups. The majority of Muslims are circumcised at birth, however, some tribal groups circumcise during adolescence as a rite of passage into manhood, and other groups do not circumcise at all. [50] Because of the wide range of beliefs it is important to take into account the cultural acceptability of male circumcision, especially neonatal male circumcision between the different religious and ethnic groups. The fact that some groups use the procedure as a rite of passage into manhood means that mandating neonatal male circumcision could cause anger because the procedure would be impeding on years of cultural tradition. For this reason mandating neonatal male circumcision may be more accepted among currently non-circumcising groups as opposed to currently circumcising groups who use the procedure as a rite of initiation.[51] "In countries such as South Africa, for example, most men are not circumcised, but certain subpopulations, including the Xhosa ethnic group, practice male circumcision of boys as a rite of passage into manhood. Many South Africans frown on the practice, and after several young Xhosa boys died from circumcision-related complications, then-President Thabo Mbeki signed a bill banning (with some religious and medical exceptions) circumcision in boys under 16 years of age. Some fear that the deaths associated with traditional circumcision have prevented expansion of the program in South Africa, but others argue that offering clean, safe medical circumcision to these communities could be lifesaving."[52] Therefore, with proper education the non-circumcising groups could be shown that the medical benefits of male circumcision are in the best interest of their people.

Public health researchers believe there are deeper reasons for why some African governments and their people are skeptical about mandating male circumcision. "Some speculate that Africa's colonist history has left these leaders with lingering suspicions about possible oppression, which have long taken the form of 'deep denial regarding HIV treatment and prevention in certain regions of Africa.' Others reference the dark history of surgical interventions deployed in the name of public health, citing the Indian sterilization camps of the 1970s. All agree that implementation of male circumcision on a national level will require in-country champions and strong political will to succeed."[53] However, since the release of the three clinical trials in 2005 and 2007 showing that male circumcision reduces the risk of acquiring HIV by 60% and the endorsement of male circumcision by the WHO, UNAIDS and PEPFAR there has been a cultural shift toward men pursuing circumcision. [54] With this cultural shift also come serious concerns about adequate medical facilities, proper training of clinicians, funding and misinformation that male circumcision guarantees full protection from HIV.

Although male circumcision is an integral part of decreasing the spread of HIV/AIDS it cannot be used as the only strategy to decrease the spread of the HIV virus. The benefits and risks of male circumcision need to be clearly communicated to both those being circumcised and to the community. It must be completely understood that the benefits of male circumcision are relative and are in no way 100% effective.[55] Communication and education are essential components if male circumcision is going to be an effective way of reducing the spread of HIV/AIDS in sub-Saharan Africa. Male circumcision also needs to be coupled with other known preventative techniques to maximize its effectiveness. The procedure presents an opportunity for different religious and ethnic groups to cooperate in designing programs to educate the public on HIV/AIDS that would encompass all strategies used to decrease the spread of the HIV virus. Such strategies include education about abstinence, the use of condoms, limiting sexual partners, mutual monogamy, testing for HIV, and the use of anti-retrovirals. [56]

There are also some risks regarding male circumcision that need to be communicated. In neonatal male circumcision the foreskin of male infants is not completely developed and neonatal male circumcision interrupts the natural separation of the foreskin from the glans possibly interfering with penile development. [57] In adolescent and adult male circumcision there are some risks such as bleeding, infections and on rare occasions secondary surgery might be needed.[58] However, these risks are rare if the physicians performing this procedure are well-trained. Another risk is that circumcision can reduce sensitivity, and then men might be less likely to use condoms and could become infected with HIV sooner rather than later. [59] In addition, if neonatal circumcision were made mandatory it would be more than a decade before any tangible results were seen. [60] It is also possible that other more effective HIV preventative measures, such as topical microbicides or an HIV vaccine could be discovered before a male circumcised at birth would reach sexual maturity.[61] Under the present circumstances, the possible risks and disadvantages of male circumcision are clearly outweighed by the benefit of decreasing HIV infections in sub-Saharan Africa.

Promoting male circumcision as a preventative technique for decreasing the risk of acquiring the HIV virus could also lead to an increase in the procedure being performed in non-favorable conditions. It is possible that the perception of the effectiveness of the procedure could cause it to be performed in non-sterile conditions with non-sterile equipment, possibly by a traditional healer, which could lead to the spread of infections, uncontrolled bleeding, permanent injury or even death. [62] It is also cheaper for male circumcision to be performed by a traditional healer. A traditional healer in Tanzania will perform a circumcision for 2500 shillings (approximately 2 US dollars) as opposed to the 10,000 shillings (approximately 8 US dollars) it costs for a circumcision to be performed in a hospital setting. In addition, many sub-Saharan African tribes perform the procedure as a rite of passage into manhood. During this initiation a respected male elder is chosen by the tribe to circumcise the boys. The young males are then lined up and are circumcised one after the other with the same knife which is only washed with water between each circumcision. This is a very dangerous practice because the procedure is not performed in a sterile environment and the knife is not sterilized between each circumcision causing the possibility of spreading the HIV virus from male to male if one of the males is HIV positive. [63] In addition, to demonstrate their new sense of manhood, recently circumcised boys are encouraged to engage in sexual intercourse soon after the rite of passage. Encouraging these young men to engage in sexual intercourse so soon after the circumcision increases the chances of HIV infection because the penis has not had sufficient time to heal. [64] These concerns could be minimized by incorporating the traditional healers into the medical establishment and educating them on the need to use sterile techniques and the importance of responsible sexual relations.

Recently the W.H.O. and UNAIDS officially recommended male circumcision as a way to prevent the heterosexual spread of the HIV virus. [65] Despite the risks and challenges that surround mandatory male circumcision medically, culturally and religiously, the potential benefits appear to make the risks and challenges justified and reasonable. The relative minimal cost of the procedure, the effect the procedure could have on decreasing the spread of HIV/AIDS in sub-Saharan Africa and the saving of millions of lives would medically and culturally justify mandating male circumcision as a public health policy. However, mandating male circumcision presents major problems. First, how does one define "mandatory"? In general, if something is mandatory it means that all applicable people must do it. For example, if car insurance is mandated for all drivers, then all drivers must have car insurance before being permitted to drive or face fines or imprisonment. If a particular vaccination is mandated in order to attend a public school, then every student must undergo this vaccination or they cannot attend the school. If this is the meaning of "mandatory," then the sub-Saharan countries bear the moral responsibility to ensure that every male in this area undergoes circumcision. This creates numerous medical, legal and ethical issues. To overcome the practical issues of how these resource-strapped countries could enforce a mandate in the traditional sense of the word and what punitive measures would be enacted if men objected; a clear meaning of the term "mandatory" must be established. For the purpose of this paper, the meaning of "mandatory male circumcision" will entail making this procedure universally available, economically affordable, medically safe and then aggressively promoting it as a vital part of a comprehensive plan to fight the spread of HIV. The next question is whether making male circumcision mandatory in sub-Saharan Africa is ethically acceptable.

Ethical Analysis

The ethical controversy surrounding the debate over mandatory male circumcision in sub-Saharan Africa has taken on a sense of urgency because every day we delay in implementing male circumcision as part of a comprehensive health policy more individuals will become HIV infected and more deaths will occur. Almost three-quarters of all adult and child deaths due to AIDS in 2007 occurred in sub-Saharan Africa: 1.5 million [1.3 million – 1.7 million] of the global total of 2.0 million [1.8 million – 2.3 million]. Overall, sub-Saharan Africa is home to an estimated 22.0 million [20.5 million – 23.6 million] adults and children infected with HIV. [66] Proponents of mandatory neonatal male circumcision argue that the protective effect is greater when circumcision takes place early in a man's life. In addition, at an early age circumcision is less costly, has less medical risks, and lessens the problem of risk compensation. [67] Proponents of adolescent and adult male circumcision argue that over the next twenty years the WHO and UNAIDS estimate that male circumcision could prevent 5.7 million more infections, could prevent 300,000 deaths in the next ten years and could avert more than 3 million deaths over the next 20 years. [68] Besides saving lives, male circumcision is cost effective. The cost effectiveness is important because it is not just limited to preventing HIV infections and the associated cost of treatment but the economic benefits gained for entire countries by maintaining the health of the most productive age groups of their populations. [69] Opponents argue that mandatory neonatal male circumcision violates the child's right of autonomy and in particular, informed consent. The removal of healthy foreskin that entails medical risks without the informed consent of the individual does more harm than good. In addition, both neonatal and adolescent/adult circumcision also violates the principle of nonmaleficence. For neonates, less invasive and more effective methods of HIV prevention such as topical microbicides, pre-exposure prophylactic drugs and even a vaccine could be discovered before the individual reaches sexual maturity.[70] For adolescents and adults, there are questions about male circumcision reducing sensitivity and function as well as the safety of the procedure if performed by untrained health care professionals.[71] Besides cultural and religious questions surrounding male circumcision, opponents also raise the practical issues of how it would be mandated and what punitive measures would result. To determine if male circumcision is ethical, and to address the ambiguities and unresolved issues surrounding this controversy, the traditional ethical principle of double effect will be applied to this situation.

Society in general has always recognized that, in our complex world, there is the possibility that we may be faced with a situation which has two consequences: one good and the other evil. The time-honored ethical principle that has been applied to these situations is called the principle of double effect. As the name implies it refers to one action with two effects. One effect is intended and morally good; the other is unintended and morally evil. As an ethical principle, double effect was never intended to be an inflexible rule or mathematical formula, but rather an efficient guide to prudent moral judgment in solving difficult moral dilemmas. [72] Historically, many ethicists believe the premises for the principle can be found in the writings of Thomas Aquinas in his famous explanation of the lawful killing of another in self-defense in the *Summa Theologicae II*, q. 64, a. 7c. However, other ethicists argue that the four conditions of the principle were not finally formulated until the mid-nineteenth century by Jean Pierre Gury. [73] The principle of double effect specifies four conditions that must be fulfilled for an action with both a good and a bad effect to be morally justified.

1) The action, considered by itself and independently of its effects, must not be morally evil. The object of the action must be good or indifferent.
2) The evil effect must not be the means of producing the good effect.
3) The evil effect is sincerely not intended, but merely tolerated.
4) There must be a proportionate reason for performing the action, in spite of its evil consequence. [74]

It should be noted that a number of ethicists known as proportionalists have argued that the first three conditions of the principle of double effect are incidental to the principle, and that in reality it is reducible to the fourth condition of proportionate reason. While this is a legitimate argument, it is not the purpose of this article to reopen the controversy on the validity of the first three conditions. This article will remain within the framework of the four conditions of the principle of double effect and apply these conditions to mandatory male circumcision as an effective method of prevention against HIV infection in sub-Saharan Africa

The principle of double effect is applicable to the issue of mandatory male circumcision in sub-Saharan Africa because it has two effects: one good and the other evil. The good effect is that three recent randomized clinical trials in Africa have shown that male circumcision reduces the risk of heterosexually acquired HIV infection in men by approximately 51% to 60%. The evil effect concerns the principles of autonomy and nonmaleficence. The removal of a neonate's healthy foreskin without his informed consent could violate the principle of autonomy. It is true that neonates do not five informed consent; parents consent for them. However, because of the serious nature of this decision, many argue the decision should be made at a time when the child can give at least informed assent. It also violates the principle of nonmaleficence in that for neonates less invasive and more effective methods of HIV prevention may be discovered before the child reaches sexual maturity. For adolescents and adults it may reduce sensitivity and function and complications may increase depending on the experience and expertise of those performing the surgery. Weighing the good and evil effects, mandatory male circumcision in sub-Saharan Africa is ethically justified because it meets the four conditions of the principle of double effect.

The *first condition* allows for male circumcision because the action in and of itself is good, in that clinical research studies have proven that it can effectively reduce male heterosexual HIV infection by 60%. In the South African study Auvert et al., argue that "male circumcision provides a degree of protection against acquiring HIV infection, equivalent to what a vaccine of high efficacy would have achieved."[75] In addition, new studies have shown that circumcision provides increased protection against HPV, herpes simplex virus, syphilis and chancroid.[76] The immediate goal is not to deny individuals their basic human right of autonomy or to inflict unnecessary harm. Rather, the direct goal is to decrease the HIV infection rate in sub-Saharan Africa and to save innocent human lives.

The *second condition* allows for mandating male circumcision because the good effect is not produced by means of the evil effect. Male circumcision provides a degree of protection against acquiring HIV infection not only for the individual but for others in society. A study in 2005 of the medical records of 300 Ugandan couples estimated that circumcised males infected with HIV were about 30% less likely to transmit it to their female partners.[77] Ethically, from a public health perspective, if we allow and in some cases require childhood vaccinations such as those against diphtheria, tetanus, polio, etc. for the good of the child and society as a whole, then how is neonatal male circumcision, which can offer long-term vaccine-level protection against HIV transmission especially for children living in a high HIV-prevalence area, any different? To exclude the use of an intervention with a potentially substantial beneficial population-level health effect violates the basic norms of ethics. And to restrict commonly accepted parental rights to choose what they believe is in their child's best interest or what society believes is in the child's best interest is also clearly unethical. We allow parents to make important medical decisions for their children on a daily basis. The decision to vaccinate a child is no different than a decision to offer a reasonable form of prevention, such as neonatal male circumcision, that is cost-efficient and has minimal medical risks. [78] For adolescents and adults, male circumcision "is the only proven medical intervention that can complement condom use and improve protection. If we had this level of data for a vaccine or a microbicide, you can bet there would be a massive push for immediate scale-up."[79] The risks of the procedure can be substantially minimized through proper education of health care professionals. It would appear that the good effect and the evil effect happen simultaneously or independent of one another.

The *third condition* is intricately tied to the second condition. The direct intention of mandating male circumcision is to provide a degree of protection to the individual male against HIV infection and to offer a degree of protection for society as a whole. Opponents of neonatal male circumcision argue that because this is the removal of healthy foreskin that can have long-term consequences and medical risks that it should be done either as an adolescent, when the child can at least give assent or as an adult when informed consent can be given. If parents are given all the facts regarding the risks and benefits of neonatal male circumcision and understand that it not only can offer additional protection for their son against HIV infection in the future but also reduces the incidence of penile cancer, urinary tract infections and reduces the susceptibility to certain sexually transmitted diseases, then their intention to allow for the procedure is good. [80] In addition, male circumcision at a later age can cause more serious medical complications. The medical complications include pain, hematoma, bleeding, damage to the penis, decrease sexual pleasure or function, infection and even death. Studies have shown that these complications are rare and with trained medical professionals the risk of complications is reduced to an even greater degree. In addition, recent studies have also shown that "circumcision status was not associated with any sexual dysfunction, or with specific sexual dysfunctions (premature ejaculation, pain during intercourse, erectile dysfunction, inability to reach organism, or lack of please during intercourse)."[81] Medical complications increase when circumcisions are performed by traditional healers with a ritual knife. [82] Training medical personnel to perform adolescent and adult circumcisions is crucial. Various programs have been established in sub-Saharan Africa to train nurses and midwives to back-up qualified surgeons. In addition, in March 2009, the WHO, in conjunction with UNAIDS, the AIDS Vaccine Advocacy Coalition, Family Health International and several American and British schools of public health, created a website [malecircumcision.org] to promote safe male circumcisions. The primary purpose of the site is to gather scientific studies, policy documents and news articles to educate against popular myths. However, the site has a handbook demonstrating surgical techniques but is not a handbook for training traditional healers or even surgeons. Instead, it could be used at small missionary hospitals in rural areas, demonstrating new methods they could investigate and teaching that men need not just surgery but pain management, sexual education and condoms. [83] The direct intention of male circumcision for neonates, adolescents and adults is to provide a reasonable protection against HIV infection for the individual and society as whole, not to violate a person's autonomy or to cause harm.

Finally, the argument for the ethical justification of mandatory male circumcision in sub-Saharan Africa by the principle of double effect focuses on whether there is a proportionately grave reason for allowing the foreseen but unintended possible consequences. To make this determination "the probability and magnitude of the good (intended) effects will have to be balanced against the probability and magnitude of the bad effects in order to determine" if there is a proportionate reason for allowing male circumcision under these circumstances. [84] Proportionate reason is the linchpin that holds this complex moral principle together.

Proportionate reason refers to a specific value and its relation to all elements (including premoral evils) in the action. [85] The specific value in mandating male circumcision is allowing for a significant degree of protection against acquiring HIV infection for both the individual and society as a whole. The premoral evil, which can come about by trying to achieve this value, is the foreseen but unintended possibility of denying the individual neonate the right of informed consent and the possibility of other less invasive and more effective methods of HIV prevention that could be discovered in the near future. For adolescents and adults it may reduce sensitivity and function and could cause other complications due to the lack of experience and expertise of those performing the surgery. The ethical question is: does the value of reducing the risk of HIV acquisition by 60% in males in sub-Saharan Africa outweigh the premoral evil of the potential violation of the principle of informed consent and the potential for the discovery of less invasive and harmful alternative methods of prevention for neonates and the possibility of potential harm to adolescents and adults? To determine if a proper relationship exists between the specific value and the other elements of the act, ethicist Richard McCormick proposes three criteria for the establishment of proportionate reason:

1) The means used will not cause more harm than necessary to achieve the value.
2) No less harmful way exists to protect the value.
3) The means used to achieve the value will not undermine it.[86]

The application of McCormick's criteria to mandatory male circumcision in sub-Saharan Africa supports the argument that there is a proportionate reason for the state to mandate it neonates and for adolescents and adults. *First*, male circumcision, when performed by certified medical personnel who follow established standards of treatment and care, is a safe procedure, cost-effective and presents minimal risks. Complications depend on a host of factors including training of medical personnel doing the procedure, availability of instruments, the level of sterility under which the surgery is done and resource-limited settings. In large studies of neonatal circumcision in the United States, complications range from 0.2% to 2.0%. [87] The risk of neonatal male circumcision in Africa is also low according to most studies. For adolescents and men, recent African studies, especially in Kenya, showed complication rates ranged from 1.7% to 3.6% among HIV-negative men in clinical trials. Most of these complications were minor—pain and bleeding—but higher complication rates have been reported outside clinical settings. These were attributed to inadequate sterilization procedures and surgical instruments. [88] Some complications include: lacerations, skin loss, skin bridges, chordee, meatitis, stenosis, urinary retention, glans necrosis, penile loss, hemorrhage, sepsis, gangrene, meningitis, possible reduction of sexual preasure and function, and even death. [89] Proponents argue that besides reducing the risk of HIV infection by 60% it also protects against the development of phimosis (narrow foreskin opening), balanitis (infected foreskin), paraphimosis (a problem seen in elderly men requiring intermittent or chronic bladder catheterization) and appears to reduce the risk of cervical cancer among female partners of circumcised men. Other medical benefits include reduced incidence of penile cancer, reduced incidence of urinary tract infections and possible reduced susceptibility to certain sexually transmitted diseases including chancroid, herpes and syphilis. [90] Safe male circumcision as part of a comprehensive plan has a critical role to play in decreasing the HIV infection rate in sub-Saharan Africa. The risks associated with male circumcision in sub-Saharan Africa are not only outweighed by the benefits of the potential lives saved, but the means used do not cause more harm than necessary if proper safeguards are in place.

Second, at the present time sub-Saharan Africa is being devastated by the AIDS pandemic. Infections rates are rising, the potential for an AIDS vaccine seems years in the future, if ever, and the potential of male microbicides is only a possibility. Recently, studies have shown that male circumcision provides a degree of protection against acquiring HIV infection that is equivalent to what a vaccine of high efficacy could achieve. Presently, there is no less harmful procedure or vaccine available that affords males this level of prevention. "Ultimately, even if an HIV preventive vaccine or microbicide were to be developed, they are unlikely to be 100% effective." [91] It is true that neonatal male circumcision would not have an immediate impact on HIV transmission. The full impact on prevalence and deaths will be apparent about ten to fifteen years later. [92] However, the impact is substantial. By contrast, adolescent and adult male circumcision would have an immediate impact.

Examining this issue from both a risk-benefit analysis and a cost-benefit analysis would dictate that the savings of lives and the impact on the economies of sub-Saharan countries would be formidable. Adolescent and adult male circumcision would eliminate many of the ethical concerns surrounding issues of informed consent and autonomy. However, issues of cost, safety, and the problem of risk compensation make neonatal male circumcision more practical and beneficial. The combination of neonatal and adolescent/adult male circumcision would provide the most comprehensive approach. What has to be stressed is that male circumcision is not a "magic bullet" as will likely be true for future microbicides or AIDS vaccines. Male circumcision confers only partial protection and must be viewed as one part of a comprehensive HIV prevention plan.

Third, mandatory male circumcision does not undermine the value of human life. One can argue convincingly that the intention of this procedure is to decrease the HIV infection rate and ultimately, to save human lives. Mandatory male circumcision, as part of a comprehensive HIV prevention plan, will not only save the lives of those who are circumcised but "will indirectly protect women and, therefore children from HIV infection because if men are less susceptible to HIV acquisition, women will be less exposed." [93] As a result of mandating male circumcision, there is the possibility that negative effects could arise such as fear of stigma, risk compensation, interfering with traditional rites of passage and the possibility of portraying African men as being vectors for this disease. These negative effects can be overcome by establishing legal statutes to protect individual's rights and by effective education of the public. With a comprehensive education plan, the care and support of those in the medical community, and the financial support of private foundations and NGO's, mandatory male circumcision would not only decrease HIV infection rates and save potential lives but would raise awareness about HIV/AIDS and make us more sensitive to the basic dignity and respect all people with this disease deserve. Male circumcision is an effective therapy that has minimal risks, is cost efficient and will save human lives. To deny individuals access to this effective therapy is to deny them the dignity and respect all persons deserve. Male circumcision not only passes the four conditions of the principle of double effect but demonstrates clearly that the greater good is promoted in spite of the potential evil consequences.

GUIDELINES

Before mandatory male circumcision can become a public policy in sub-Saharan African countries there needs to be extensive education and counseling done to inform people about the procedure and why this new initiative is necessary to combat HIV/AIDS. "Governments considering whether and how to enhance the availability of safe male circumcision services will need to consider how to contextualize male circumcision within comprehensive HIV prevention programming and what prominence to afford male circumcision services in relation to other health services given current financial and human resource constraints." [94] Emphasis has to be put on the fact that although male circumcision may reduce HIV infection, it is not a cure all. Male circumcision, whether it is neonatal, adolescent or adult, is only one part of a HIV/AIDS strategy that must entail a comprehensive approach to prevention. Extensive education is needed to explain not only the scientific and medical evidence about this procedure but to address religious and cultural concerns that may also present problems concerning mandatory male circumcision. This education must have a solid evidence-base in order to present the facts in a transparent way so that any suspicion about mandating this procedure is alleviated. To further ensure the trust and effectiveness of this initiative it must be grounded in respect for persons, social justice, human rights and community values.[95] Included in this education should be cost-effectiveness data that would provide further evidence to support the health benefits of male circumcision. The following guidelines should be implemented before mandatory male circumcision becomes part of a comprehensive HIV/AIDS prevention strategy:

1. Initiation of research studies to investigate whether male circumcision will be accepted, and whether traditional circumcision rites could be incorporated into a general health policy.[96]
2. Initiation of operation studies to compare different technologies and health-system approaches to scale up male circumcision and establish safety and cost effectiveness. A cost analysis must be provided that is not limited to preventing HIV infections and the associated costs of treatment, but the economic benefits gained for entire countries by maintaining the health of the most productive age-groups. One such program is the Male Circumcision Partnership, which is launching a massive scale-up voluntary male circumcision program in Swaziland and Zambia. The Partnership is supported by a five-year, $50 million

grant from the Bill and Melinda Gates Foundation to Population Services International (PSI). [97]

3. Initiation of further modeling studies to determine if over time women could benefit from an effect similar to the herd immunity seen with mass immunizations. Further clinical trials should also be initiated to determine if male circumcision might also directly protect against male-to-female transmission of HIV. A trial to test this hypothesis in currently underway in Uganda, with preliminary results expected in 2009.[98]
4. Coordinate scientific studies, policy documents and news articles as part of an education and prevention package that would help to fight popular myths about male circumcision, like the one that circumcision is 100% protective so men can stop using condoms. This type of educational material would be part of a comprehensive education/ prevention package.[99]
5. Regional and country task forces on male circumcision and HIV prevention should be established, or where already established the membership should be enlarged to ensure inclusivity, to coordinate and direct efforts around male circumcision as a potential HIV prevention tool. There should be broad consultation and inclusion of religious and ethnic groups (including traditional healers) because male circumcision carries major religious, social and cultural meanings. Women should also be included in this process. [100]
6. Development of national comprehensive strategic plans for mandatory male circumcision which include assessment of current capacity, development of a single sentinel surveillance and reporting mechanism, clear guidelines for basic standards of care, measurable targets, cost analyses, specific assessment of internal ethnic group practices and future potential increases in demand.[101]
7. Development of national and regional "tool kits" for ministries of health that include manuals and modules of training of practitioners as well as for training of trainers. [102]
8. Implementation of communication strategies that not only provide information and education about the effectiveness of male circumcision and the risks and benefits of the procedure; but further develop new innovative prevention messages. This education must be sensitive to the concern that male circumcision does not lead to an increase in the practice of female genital mutilation. Emphasis must be given to the fact that female genital mutilation is not the same as

male circumcision and has no medical benefits; instead it may promote disease transmission and acquisition. [103]

9. Establish competency levels and requirements for certification that ensure that male circumcisions are performed by individuals with primary-care surgical skills in facilities that can guarantee the sterility of the procedure, appropriate wound dressing to ensure low rates of adverse events and minimization of pain through proper anesthesia. This would include training midwives and nurses to back-up the qualified surgeons. Handbooks demonstrating surgical techniques, like the one established by the W.H.O. and UNAIDS on their new website www.malecircumcision.org would be beneficial to medical professionals.[104]
10. Implementation of education and prevention programs emphasizing the fact that male circumcision combined with other prevention strategies such as viral load suppression, abstinence, mutual mono-gamy, reduction in the number of sexual partners', correct and consistent use of male and female condoms, etc., could dramatically reduce the infection rates for both men and women. One example of an innovative education/prevention program is a musical group called "Circ Squad" who got circumcised and made a music video about the issue for educational purposes. Other examples would be governments running television and radio campaigns to encourage men to visit clinics for safe circumcision procedures as is being done in Botswana.[105]
11. Facilitate equitable access of low-cost circumcisions that minimize potential burdens on already fragile health systems by establishing cooperation and commitments from multilateral, bilateral and government agencies, along with non-governmental organizations. At the present time because of the costs male circumcision is limited to middle and upper-class men. This life-saving strategy needs to be affordable and safely available to relevant populations.

CONCLUSION

Just as a combination of treatment for individuals with HIV is more effective than a single drug therapy, one can argue that a combination strategy for sexually active people will be more effective than reliance on one single HIV prevention method. [106] At the present time, mandating male circumcision as the cornerstone of a comprehensive HIV prevention strategy offers the best hope for sub-Saharan Africa. News that Merck's much-heralded HIV vaccine, which was in clinical trials, had been halted in September 2007 because the vaccine failed to prevent infection or reduce the severity of infection has been a devastating blow to all concerned with AIDS research. [107] This latest vaccine failure highlights the importance of male circumcision as a reasonable alternative to decrease the spread of HIV in sub-Saharan Africa. Even though male circumcision will not afford complete HIV protection it can become an integral part of a comprehensive HIV prevention strategy that will decrease HIV infections and help to save lives in the future. If the guidelines proposed are implemented and safeguards are assured, then mandatory male circumcision is not only medically necessary, it is ethically imperative. If the protection and preservation of human life is a priority in sub-Saharan Africa, then it is time to initiate mandatory male circumcision before it is too late for the most vulnerable. This must become an immediate priority, because human lives hang in the balance.

REFERENCES

[1] UNAIDS, "Report on the Global AIDS Epidemic-Executive Summary." July 2008; 1-31. Geneva, *UNAIDS.* *http://data.unaids.org/pub/GlobalReport/2008/JC1511_GR08_Executive Summary_en.pdf.*

[2] *UNAIDS*, 1-31.

[3] *UNAIDS*, 1-31.

[4] *UNAIDS*, 1-31.

[5] Weiss HA et al: Male Circumcision and Risk of HIV Infection in sub-Saharan Africa: A Systematic Review and Meta-Analysis. *AIDS,* 2000; 14: 2361-2370.

[6] Bailey RC et al: Male Circumcision for HIV Prevention in Young Men in Kisumu, Kenya: A Randomized Controlled Trial. *The Lancet,* February 24, 2007; 369: 643-656.

[7] Gray RH et al: Male Circumcision for HIV Prevention in Men in Raki, Uganda: A Randomized Trial. *The Lancet,* February 24, 2007; 369: 657-666.

[8] Auvert B et al: Randomized, Controlled Intervention Trial of Male Circumcision for Reduction of HIV Infection Risk: The ANRS 1265 Trial. *PLoS Medicine,* November 2005; 11: 1112-1122.

[9] MacInnis L: U.N. Recommends Circumcision to Reduce Male Heterosexual HIV Acquisition. *Medscape-Reuters Health Information,* March 29, 2007; 1-2. *http://www.medscape.com/viewarticle/554307_print*
[10] Centers for Disease Control: Male Circumcision and Risk of HIV Transmission: Implication for the United States. CDC HIV/AIDS Science Facts, March 2007; 1-6. *http://www.cdc.gov/hiv* See also, Szabo R, Short RV: How Does Male Circumcision Protect Against HIV Infection? *British Medical Journal,* June 10, 2000; 320: 1592-4.
[11] Centers for Disease Control: Male Circumcision and Risk of HIV Transmission: Implication for the United States. *CDC HIV/AIDS* Science Facts, March 2007; 1-6.
[12] Clark P: To Circumcise or Not to Circumcise. *Health Progress,* September-October 2006; 87: 30-39.
[13] *Auvert*, 1112-1122.
[14] *Clark*, 31.
[15] *Clark*, 30-39.
[16] *Clark*, 30-39.
[17] *Clark*, 30-39.
[18] *Bailey* et al., 643-656 and Gray et al., 657-666.
[19] Morris, B: Clinical Innovations Introduces New Circumcision Device. Clinical Innovations, Murray, Utah. June 8, 2009. *www.accucirc.com*
[20] Morris, 1.
[21] Holman J et al: Adult Circumcision. *American Academy of Family Physician,* March 2000: 1-5. *http://www.aafp.org/afp/990315ap/1514.html*
[22] *MacInnis*, 2.
[23] *MacInnis*, 1-2.
[24] *MacInnis*, 1-2.
[25] *MacInnis*, 1-2.
[26] *MacInnis*, 1-2.
[27] *MacInnis*, 1-2.
[28] *Weiss* at al., 2361-2370; Bailey et al., 643-656; and Gray et al., 657-666.
[29] *Bailey* et al., 643-656.
[30] *Gray* et al., 657-666.
[31] *Auvert* et al., 1112-1122.
[32] *MacInnis*, 1-2.
[33] Rennie S et al: Male Circumcision and HIV Prevention: Ethical, Medical and Public Health Tradeoffs in Low-Income Countries. *Journal of*

Medical Ethics, 2007; 33: 357-361. *http://jme.bmj.com/chi/content/full/33/6/357*

[34] *Rennie* et al., 357-361.

[35] Tobian A. et al: Male Circumcision for the Prevention of HSV-2 and HPV infections and Syphilis. *New England Journal of Medicine,* 2009; 360: 1298-1309.

[36] Auvert B. et al: Effect of Male Circumcision on the Prevalence of High-Risk HPV in Young Men: Results of a Randomized Controlled Trial Conducted in Orange Farm, South Africa. *Journal of Infectious Disease,* 2009; 199: 14-19.

[37] Austin P. Is Neonatal Circumcision Clinically Beneficial? Argument in Favor. *Nature Clinical Practice Urology*, 2008; 6: 16-17.

[38] Warner L. et al: Male Circumcision and the Risk of HIV Infection Among Heterosexual African American Men Attending Baltimore Sexually Transmitted Disease Clinics. *The Journal of Infectious Diseases,* 2009; 199: 59-65.

[39] *Tobin* et al., 1298-1309.

[40] *Bailey* et al., 643-656.

[41] Auvert et al, Radomized, *Controlled Intervention Trial of Male Circumcision for Reduction of HIV Infection Risk*, 1112-1122.

[42] Katz I. and Wright A: Circumcision—A Surgical Strategy for HIV Prevention in Africa." *New England Journal of Medicine*, 2008; 359: 2412-2415.

[43] *Katz*, 2412-2415.

[44] Seppa N: Men Line Up For Circumcision in Africa. *Science News,* 2009; 175: 14. *http://www.sciencenews.org/view/generic/id/39048/title/Men_line_up_for_circumcision...*

[45] *Rennie* et al., 357-361.

[46] *UNAIDS*, 1-31.

[47] Sawires S et al: Male Circumcision and HIV/AIDS: Challenges and Opportunities. *The Lancet,* February 24, 2007; 369: 708-712.

[48] *Rennie* et al., 357-361.

[49] Auvert et al, Radomized, *Controlled Intervention Trial of Male Circumcision for Reduction of HIV Infection Risk,* 1112-1122.

[50] Auvert et al, Radomized, *Controlled Intervention Trial of Male Circumcision for Reduction of HIV Infection Risk*, 1112-1122.

[51] *Rennie* et al., 357-361.

[52] *Katz*, 2412-2415.

[53] *Katz*, 2412-2415.
[54] *Seppa*, 14.
[55] *Sawires*, 708-712.
[56] *Sawires*, 708-712.
[57] *Clark*, 30-39.
[58] Smith S: To Cut Or Not To Cut: Circumcision May Help Fight AIDS in Africa, But In the US, the Medical Argument is Iffy. *Boston Globe,* October 16, 2006. *http://www.boston.com/yourlife/health/diseases/articles/2006/10/16/t0_cut_or_not_to_cut...*
[59] Keeton C: Foreskin or No Foreskin, Men Get It On, Says Study. *The Times*, April 26, 2009. *http://www.thetimes.co.za/printedition/Article3.aspx?id=987696*
[60] *Rennie* et al., 357-361.
[61] *Rennie* et al., 357-361.
[62] Auvert et al., *Radomized, Controlled Intervention Trial of Male Circumcision for Reduction of HIV Infection Risk,* 1112-1122.
[63] *Rennie* et al., 357-361.
[64] *Rennie* et al., 357-361.
[65] *MacInnis*, 1-2.
[66] *UNAIDS*, 1-31.
[67] Rennie et al., 357-361.
[68] *MacInnis*, 1-2.
[69] *Sawires*, 708-712.
[70] *Rennie* et al., 357-361.
[71] *Keeton*, 1.
[72] Mangan J: An Historical Analysis of the Principle of Double Effect. *Theological Studies,* March 1949; 10: 30-45.
[73] Kaczor C: Double-Effect Reasoning From Jean Pierre Gury to Peter Knauer. *Theological Studies,* 1998; 59: 297-316.
[74] Kelly G: *Medico-Moral Problems.* (St. Louis): The Catholic Health Association of the United States and Canada Press; 1958.
[75] Auvert et al., *Radomized, Controlled Intervention Trial of Male Circumcision for Reduction of HIV Infection Risk*, 1112-1122.
[76] *Katz*, 2412-2415.
[77] McNeil D: Circumcision Halves HIV Risk, U.S. Agency Finds. *New York Times,* December 14, 2006; 1-4. *http://www.nytimes.com/2006/12/14/health/14hiv.html?ei=5070&en=69b72317ae0b7a&d&...*
[78] *Rennie* et al., 357-361.

[79] *Katz*, 2412-2415.

[80] *Clark*, 30-39.

[81] *Keeton*, 1.

[82] *Rennie* et al., 357-361.

[83] McNeil D: New Website Seeks to Fight Myths About Circumcision and HIV. *New York Times,* March 3, 2009. *http://www.nytimes.com/2009/03/03/health/03glob.html?_r=1&pagewanted=print*

[84] Walter J: Proportionate Reason and its Three Levels of Inquiry: Structuring the Ongoing Debate. *Louvain Studies*, Spring 1984; 10: 30-37.

[85] *Walter*, 30-37.

[86] McCormick R and Ramsey P: *Doing Evil to Achieve Good.* (Chicago): Loyola University Press; 1978.

[87] Alanis MC et al: Neonatal Circumcision: A Review of the World's Oldest and Most Controversial Operation. *Obstetrics and Gynecology Survey,* May 2004; 5: 379-395.

[88] Bailey R et al: Male Circumcision for HIV Prevention: A Prospective Study of Complications in Clinical and Traditional settings in Bungoma, Kenya. *Bull World Health Organ,* 2008; 86: 669-677.

[89] Svoboda JS: Circumcision-A Victorian Relic Lacking Ethical, Medical or legal Justification. *American Journal of Bioethics*, Spring 2003; 3: 52-54.

[90] Task Force on Circumcision, American Academy of Pediatrics: Circumcision Policy Statement. *Pediatrics,* March 1, 1999; 103: 386-396, available at *www.cirp.org/library/statements/aap1999/*: 1-18.

[91] Editor: Newer Approaches to HIV Prevention. *The Lancet,* February 24, 2007; 369: 615.

[92] Williams B et al: The Potential Impact of Male Circumcision on HIV in Sub-Saharan Africa. *PLoS Medicine,* July 2006; 3: 1032-1040.

[93] Auvert et al., Radomized, Controlled Intervention Trial of Male Circumcision for Reduction of HIV Infection Risk, 1112-1122.

[94] Topor Y: Male Circumcision and HIV. *UNAIDS Fact Sheet,* December 2006: 1-3. *http://www.unaids.org/en/HIV_factsheetepi2006/default.asp.*

[95] *Rennie* et al., 357-361.

[96] Newell ML et al: Comment: Male Circumcision to Cut HIV Risk in the General Population. *The Lancet,* February 24, 2007; 369: 617-619.

[97] Sawires, 708-712, Newell, 617-619 and Press Release. Unprecedented Scale-UP of Voluntary Male Circumcision Begins in Swaziland and

Zambia. *Yahoo Finance*, June 11, 2009. *http://finance.yahoo.com/news/Unprecedented-ScaleUp-of-prnews-15497861.html?v=1*

[98] *Editor*, 615.

[99] *McNeil*, 1-4.

[100] Sawires, 708-712 and U.N Regional Working Group on Male Circumcision: Regional Consultation on Safe Male Circumcision and HIV Prevention. *UNAIDS,* November 20-21, 2006; 1-12.

[101] Sawires, 708-712 and U.N Regional Working Group on Male Circumcision: Regional Consultation on Safe Male Circumcision and HIV Prevention. *UNAIDS,* November 20-21, 2006; 1-12.

[102] *Sawires*, 708-712.

[103] *Sawires*, 708-712.

[104] *Newell*, 617-619 and McNeil, 1-4.

[105] *Seppa*, 14.

[106] *Topor*, 1-3.

[107] Altman LK and Pollack A: Failure of Vaccine Test is Setback in AIDS Fight. *New York Times,* September 22, 2007; 1-2. *http://www.nytimes.com/2007/09/22/health/22vaccine.html?ref=health&page*

In: Bioethics: Issues and Dilemmas
Editor: Tyler N. Pace
ISBN: 978-1-61728-290-4

Chapter 2

MEDICAL ETHICS IN OBSTETRICS – A FRENCH EXPERIMENT

***Patrick Leblanc*[*1] and Pierre-Olivier Arduin*[≠2]**
[1]Centre Hospitalier de Béziers, Béziers cedex, France,
[2]Université catholique de Lyon, Draguignan, France

ABSTRACT

Medical ethics consist fundamentally in using rational knowledge to ensure the respect for human dignity. Medicine has always been in its very essence an ethical process as evinced by the ancient Hippocratic oath. Nowadays, ethical issues must be placed within the context of the changes in our way of doing things due to the extraordinary evolution of science and technology. In the field related to the beginning of life, progress in prenatal diagnosis has completely changed our medical approach to pregnancy. This has undeniably become one of the recurring themes in bioethical debates in France, alongside questions about medically assisted procreation, surrogacy and stem cell research. Indeed, France has generally been considered a pioneer in this field since its first bioethical laws were adopted in 1994. Having reviewed these laws three times in fifteen years, the French legislature has always deemed it

* Obstetrician-Gynecologist, Centre Hospitalier de Béziers, 2 rue Valentin Haüy, 34525 Béziers cedex, France, patrick.leblanc@ch-beziers.fr
≠ PhD student in Philosophy, Ethics and Moral Philosophy Department , Université catholique de Lyon, 795, Bld Joseph-Collomp, 83300 Draguignan, France, pierre-olivier.arduin@orange.fr

necessary to supervise biomedical practices so as to avoid reaching a point of no return. However, with medical ethics defending the very principle of humanity on the one hand and a bioethical arsenal of legislation on the other, may there not be a risk of losing coherence? Based on the daily practice of an obstetrician-gynecologist, the authors of this chapter raise questions about the possible conflicts of values, and possible irrational aspects they might reveal.

Introduction

The obstetrician's vocation is defined by the very etymology of his/her specialty: an obstetrician (*ob stare*) *stands facing* the mother as she is giving birth, delivering life. The fetus and the mother are privileged patients. Indeed, both the expectant mother and the expected child receive the obstetrician's utmost care, which defines the very nature of his/her mission. The obstetrician's essential concern is to accompany a child from pre-birth to birth. This very act and care for the unborn child fundamentally orients his/her medical ethics.

In the field related to the beginning of life, recent ethical issues may very well transform our "medical art" in view of the tremendous scientific and technical innovations. A few years ago, because of inevitable after-effects and prognoses of major handicaps, there was a broad consensus among both pediatricians and obstetricians to decide against any therapeutic surgery at 25 weeks of amenorrhea when confronted with late abortion. Today, new powerful tocolytics, exogenous surfactant therapy, and progress in intensive care give hope of much more reliable prognoses than in the past. High-risk patients are directed toward a tertiary level, a very specialized center, and the fetuses are no longer abandoned. Our very language has changed alongside our understanding of the pathology: we no longer talk of late-term abortion but of very preterm birth. How important words are! Our ethics are not permanently fixed then: what we used to consider harmful and wrong, such as the prolongation of life by medical means, has today become acceptable, of course with due regard to the parents. We could give many examples, including the specific care of pregnant diabetic women, which have totally altered medical prognoses...

No doubt progress made in prenatal diagnosis has drastically changed our medical approach to pregnancy. Can we however conclude that the tremendous scientific innovations of our times agree with the ethical conception of our profession? [1]. Thanks to paraclinical techniques such as ultrasounds or hormonal dosages, the obstetrician can now make diagnoses that may have very serious emotional, human, ethical, and legal consequences. Because they may lead to a termination of pregnancy, prenatal diagnoses are constantly at the core of the French bioethical debate, alongside questions pertaining to medically assisted procreation, surrogacy and stem cell research.

Indeed, France has generally been considered a pioneer in this field since its first bioethical laws were adopted in 1994. About to review them for the third time in early 2010, the French legislature has always deemed it necessary to supervise biomedical practices so as to avoid reaching a point of no return. However, with medical ethics defending the very principle of humanity on the one hand and a bioethical arsenal of legislation on the other, is coherence always possible? The possible resurgence of medical eugenics may well be at stake in today's bioethics. Following the daily practice of an obstetrician-gynecologist in the hospital of a medium-sized French town, the authors of this chapter intend firstly to give a minute description of the procedures and methods of prenatal screening, and secondly to question their relevance and ethical validity.

PRENATAL DIAGNOSIS (PND) IN TODAY'S FRENCH OBSTETRICAL PRACTICE

According to the French Public Health Code, written in accordance with the bioethical laws adopted on July 29, 2004 and reviewed on August 6, 2004, "Prenatal diagnoses refer to the medical practices that aim at detecting "in utero" a particularly serious affection in the embryo or the fetus" [2]. Such a diagnosis is made with the use of medical imaging tests (ultrasounds) and samples of amniotic liquid, trophoblasts or blood from the umbilical cord taken from the embryo or fetus in utero. Screening maternal blood (serum markers) is another way to make a PND. It is systematically proposed and performed in the very first months of pregnancy. In practice, ultrasounds and dosages of plasmic markers are the two cornerstones of PND.

Obstetricians do not work alone but are part of a network of practitioners. Hence diagnostic procedures and the course of action to take are determined in a multidisciplinary and collective way in centers associating obstetricians, sonographers, geneticists, pediatricians, and psychologists, known as "Multidisciplinary Centers for Prenatal Diagnosis" (CPDPN in French). These centers have been commissioned by the Legislature to confirm that Medical

Terminations of Pregnancy (MTP) are indicated, the approval being issued anytime during the pregnancy, once all the members have met, if there is “a high probability that the fetus is suffering from a serious and incurable illness” [2]. The Biomedicine Agency created by the decree of May 9, 2005 is the administrative organization that authorizes the centers and issues approvals for the practitioners working there. Yet, although both the work of obstetricians and their communication with the doctors in the centers are made easier by the use of videoconferencing, they remain alone when delivering the news of fetal malformation to an expectant couple. The latest report of the Biomedicine Agency states that 48 French centers issued 6,645 approvals for MTP in 2007, an increase of 10% compared with the 6,093 approvals issued in 2005 [figure 1].

	2005	2006	2007
Number of examined cases	**25 022**	**24 389**	**28 292**
Number of approvals issued for a MTP	**6 093**	**6 790**	**6 645**
Number of refusals for a MTP	**106**	**122**	**112**
Number of pregnancies with a pathology that could have led to an approval for a MTP	**406**	**402**	**475**

Figure 1. A summary of the activities of CPDPN from 2005 to 2007. in Agence de la biomédecine. *Rapport annuel et bilan des activités, CPDPN 2007*. p 282

1. Screening with Ultrasound Images

An ultrasound, both harmless and effective, with images defined to 1\10 of a millimeter, provides an accurate internal picture of the embryo and the fetus. The test has become essential to follow the morphological growth of the fetus. It is usually repeated for each term of the pregnancy: in the 3rd, 5th and 8th month. Depending on the diagnosis, the exploration can lead to other tests: X-ray, MRI or biological tests for the expectant mother, or intra-uterine samples (samples of amniotic liquid by amniocentesis, of trophoblasts by choriocentesis, or of cord blood by cordocentesis).

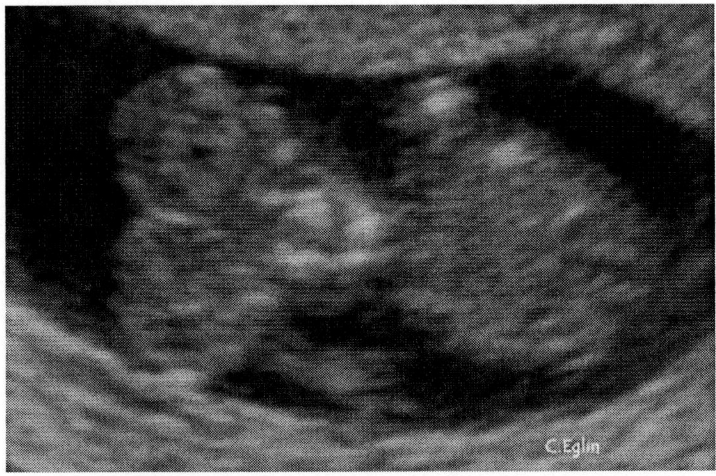

Figure 2. Acrania at twelve weeks of amenorrhea. The cerebral hemispheres, like Mickey Mouse ears, are floating in the amniotic liquid, without any calvarium. They will progressively be eroded by fetal movements and the toxic liquid. In a few weeks, acrania will evolve into anencephaly.

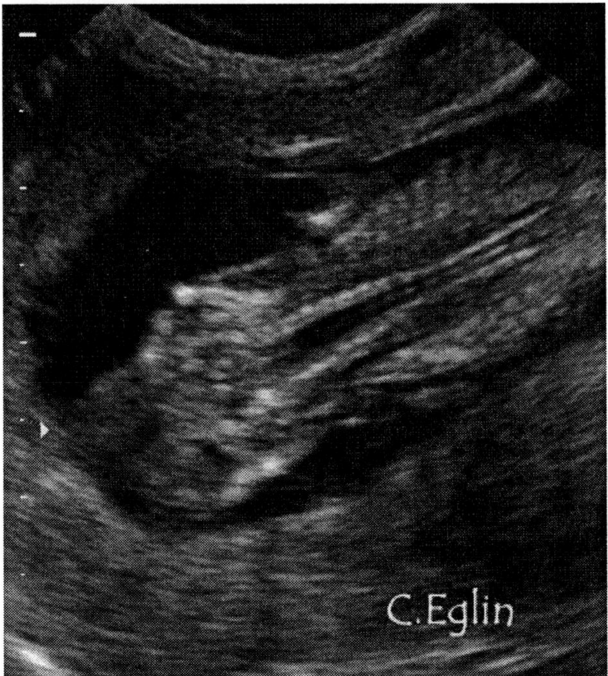

Figure 3. Anencephaly

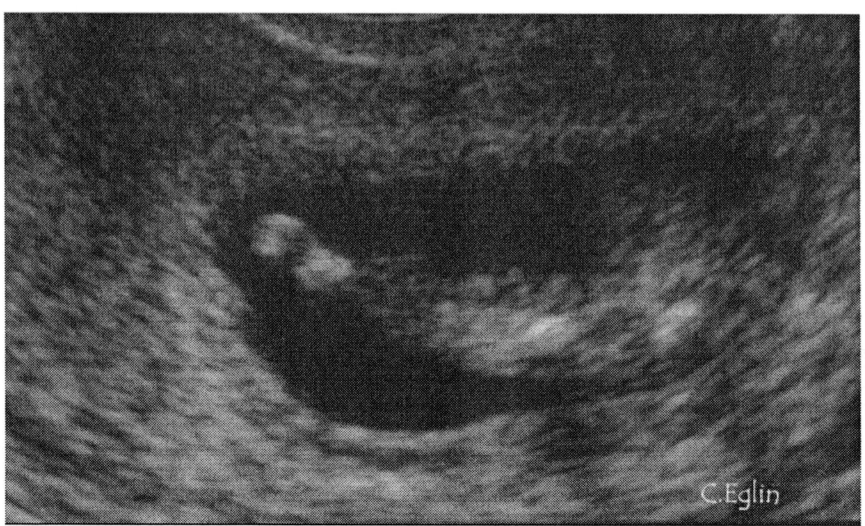

Figure 4.sirenomelia at 13 weeks of amenorrhea, the fetus has one lower limb only.

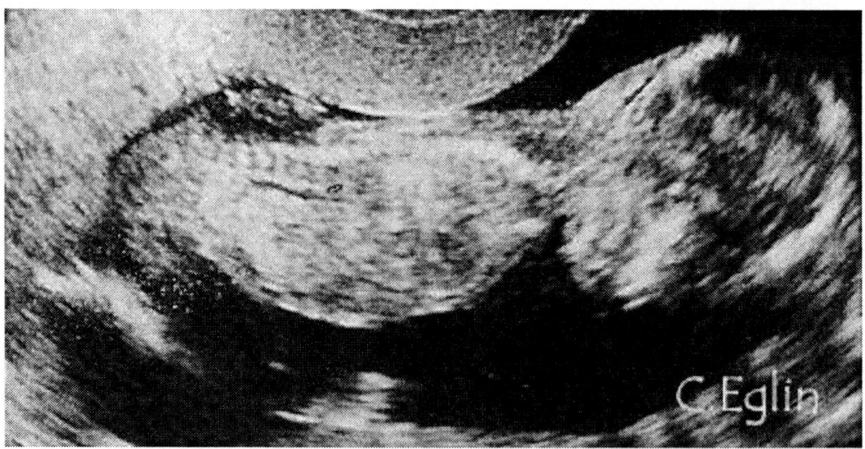

Figure 5. spina bifida detected at 12 weeks of amenorrhea with a break in the continuity of the rachis.

Some diagnoses of fetal abnormalities can be made very early and provide indisputable prognoses such as acrania [figure 2], anencephaly [figure 3], agenesia corticalis, and mermaid syndrome [figure 4] Their outcomes are known as they are constantly fatal. Others such as spina bifida [figure 5] are viable but lead to very serious complications. Fortunately abnormalities of good prognosis are diagnosed too: cleft palates [figure 6], laparoschisis are in most cases curable through surgery. It is essential to establish these diagnoses in utero in order to prepare couples to welcome their child and to make sure the new-born is taken care of in the best medical and surgical conditions.

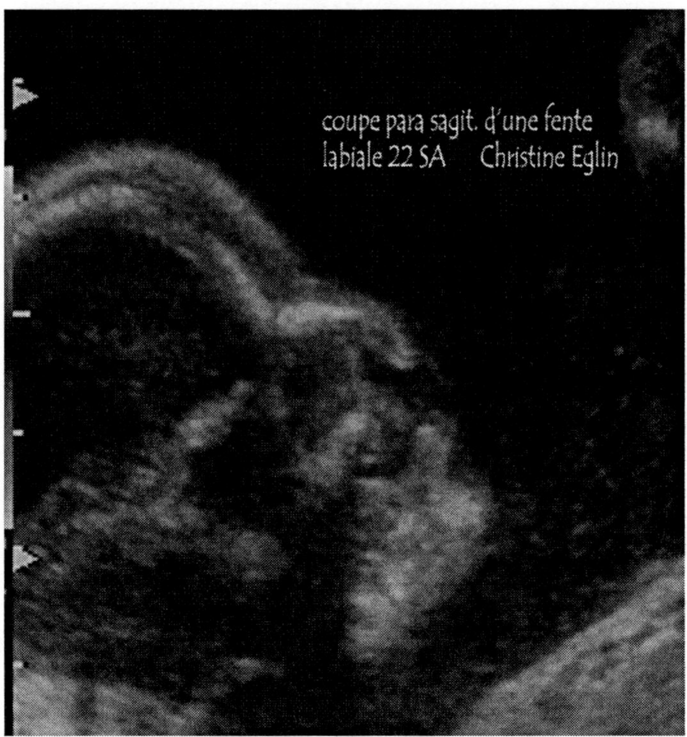

Figure 6. Cleft lip and palate with defect seen at 22 weeks of amenorrhea.

On the contrary, some diagnoses are highly contentious. Hence a partial absence of corpus callosum (a normal cerebral structure), or a supra-limited value of 1 to 2mm of an intra-cranian ventricular cavity [figure 7], induces a battery of further tests: checking the fetal karyotype, looking for infectious consequences, intra-cranian hemorrhage, coagulation difficulties... When all of these tests are negative, a doubt still remains as to some possible future mental deficiency. Such cases can be extraordinarily complex, and the experience particularly stressful for the couple as tests are repeated and specialized consultations multiplied. When in doubt, some centers will refuse a MTP. Motivated by worry and because it is legal, the couple can file a request to one or several centers till eventually the pregnancy can be terminated for medical reasons, as it remains legal whatever the age of the pregnancy. Hence, if the percentage of refusals is relatively stable, around 1.7% , of 112 cases initially refused in 2007, 40 couples were allowed to terminate the pregnancy in an other center or abroad. Thus, the centers can have opposing attitudes towards malformations of the same kind, accepting or rejecting a termination. Such heterogeneous decisions may account for reports of medical tourism [3].

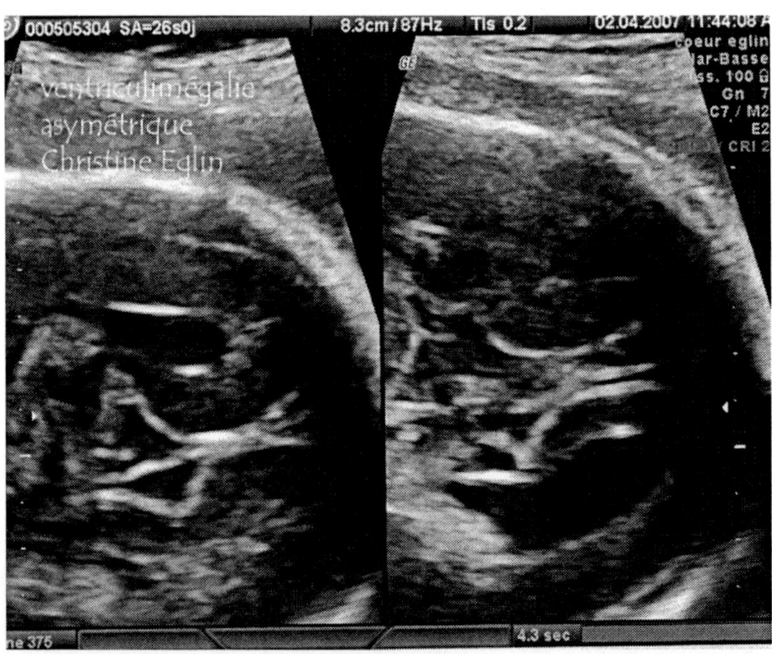

Figure 7. Asymetric intracranial lateral ventricles.

Thus there is no standardized course of action, and up to now, the Legislature has refused to list the pathologies which might be allowed for a MTP, in order not to discriminate against already-born ill subjects. Moreover, as specified by the law “a high probability that the fetus might be affected”, and not an absolute certainty, is enough to question the continuation of the pregnancy. Can the interests of the mother and of the fetus, both under the care of the obstetrician, differ so much that at a given time the decision to terminate the pregnancy can be made based on a probability, even though high? Is the presumption of innocence beneficial to the “defendant”?

In addition, fetal nuchal translucency screening done in the 1st trimester (more precisely between week 11 and week 13 and 6 days of amenorrhea) to measure the translucent space in the tissue at the back of the developing baby's neck, assesses the baby's risk of chromosomal abnormalities and in particular, Down Syndrome [figure 8]. The space can be larger in the case of increased nuchal translucency or cystic hygroma, when fluid has accumulated at the back of the neck. An amniocentesis is suggested when nuchal translucency is more than 3 mm wide or when it exceeds the 95th percentile limit, correlated with the cranio-caudal length of the embryo. 5 % of fetuses are affected.

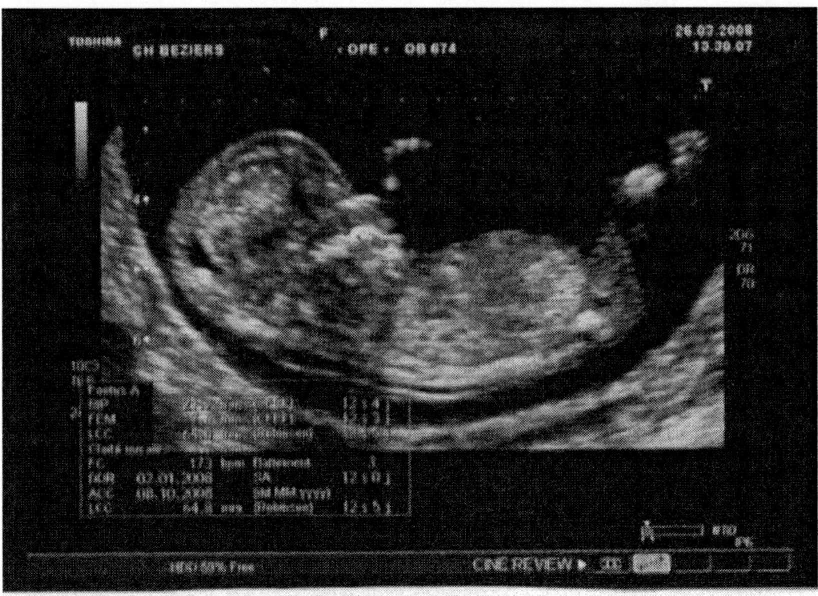

Figure 8. embryo at 12 weeks & 5 days of amenorrhea. The craniocaudal length equals 64.8 mm and nuchal translucency measures 1.7 mm.

Figure 9 highlights the link between the importance of nuchal translucency and adverse pregnancy outcome. When nuchal translucency is superior to the 95th percentile, there is a rapidly increasing risk of chromosomal abnormalities and a risk of serious non-chromosonal abnormalities (heart defects, diaphragmatic hernia...). But in most cases (especially under the 99th percentile), the new-born baby will be perfectly healthy. All this must be explained to the couple as well as the risk for chromosomal abnormalities (about 0.2%), which however cannot be ruled out even though the nuchal translucency is considered normal [4].

Measure of Nuchal translucency	Chromosomal abnormalities	Fetal death	Serious malformations	Alive and healthy new-borns
<95ième perc	0,2%	1,3%	1,6%	97%
95ième p-99ième p	3,7%	1,3%	2,5%	93%
3,5-4,4mm	21,1%	2,7%	10,0%	70%
4,5-5,4mm	33,3%	3,4%	18,5%	50%
5,5-6,4mm	50,5%	10,1%	24,4 %	30%
= or > 6,5mm	64 ,5%	19,0%	46,2%	15%

Figure 9. Link between the measure of nuchal translucency and the outcome of the pregnancy. From Souka and Nicolaïdes, 2005.

The performances of ultrasounds are a daily source of amazement, if not of wonder. It is now possible to detect a regular, fast pulsatile activity in the primitive heart tube belonging to an embryo measuring no more than 1.5 mm. Such an activity is two to three times faster than the mother's heart rate. Are we dealing with some virtual image, a thing, an animal? Could the embryo and fetus be endowed with some infrahuman status? What a major question! Although both the advocates of experimental research on the embryo and advocates of the termination of pregnancy deny any humanity to the embryo, the obstetrician is fully aware he/she is practicing medicine on a patient. All the tests induced by such a practice (ultrasound, MRI, amniocentesis...) belong to the field of "fetal medicine", a branch of our specialty or sub-specialty, and are acknowledged and reimbursed by the French Health Insurance. Besides, we do know that a fetus can feel and feel pain. Obstetricians are undoubtedly in the service of life. Yet, in their practice, they are directly confronted with the question of an ontological status - or rather the lack of it - currently given to the not-yet-born [5]. "The duplicity of its nature, or rather its uncertainty, allows, medically and legally, its protection as well as its destruction" [6]. That is a very old debate: when does life begin?... At the very first cry or on the contrary as early as conception?... Do we still have to differentiate the embryo at pre-implantation (for some, a cluster of cells) from the implanted embryo? Can there be a gradation of the human element?

Some endow the beginning of a pregnancy with humanity only if there is a parental project. Such a moral value does indeed belong to the couple and is necessary to give some destiny to the embryo. Obstetricians experience this more than once every day. They may also deplore the fact that the limit for medical terminations has been extended from 12 to 14 weeks of amenorrhea since 2001, that women more willingly accept their repetitiveness than the use of contraception, when medical terminations should be the exception, as well as the easiness of drug terminations, praised by some activists... all these factors cause confusion in the collective unconscious, trivializing and reifying pregnancy...

2. Screening with "Numbers": The Case of Trisomy 21

In 1992, only patients over the age of 38 at the time of conception were reimbursed by the French Health Insurance for an amniocentesis for fetal karyotype, owing to an increased risk of Trisomy 21 due to their age. Since then PND has gone through various stages in France.

This chromosomal abnormality can also affect, even though less frequently, the fetuses of much younger women. Leaning on the generous idea of promoting some equality among all pregnant women, whatever their ages, the authorities allowed generalized testing, reimbursed by the French Health Insurance.

Screening for Trisomy 21 of all pregnant women with maternal serum markers (MSM) was set in place nationwide by the decree of May 27, 1997. Coincidently, the new regulation took effect the year after the award for best lead actor at the Cannes Film Festival was presented jointly to Pascal Duquenne, an actor with Down Syndrome, and Daniel Auteuil, for their parts in the film "The Eighth Day".

At that time, the dosage of M.S.M. was performed between week 14 and week 17 and 6 days of amenorrhea. There are 2 even 3 associated markers: free choriogonadotropin β subunit with hCG, alpha-fetoprotein and estriol. The calculation of the risk for chromosomal abnormalities results in a fraction. The risk cut-off, calculated with software, is set at 1/250. When the risk is superior or equal to 1/250, the patient is considered to belong to an increased-risk-for-Trisomy-21 group. On the contrary, the patient who does not belong to the risk group is informed that, even though the risk is low, it still exists.

The number of cases of Trisomy 21 being diagnosed by amniocentesis among women said to be at risk is reported in figure 10. Initially set up for one year then prolonged for five years renewable, the method was designed to allow the detection of at least 60% of children with Down Syndrome. Currently, 79% of such children have been detected. Moreover, the procedure followed until recently is sequential. It includes an ultrasound performed between week 11 and week 13 and 6 days of amenorrhea, then a biochemical dosage of M.S.M. at the second trimester, between week 14 and week 17 and 6 days of amenorrhea. In 2007, for 816,500 registered deliveries, 665,054 women, that is 3/4 of all the women, were thus screened. When the risk is rated high, an amniocentesis, possible at about the 16th or 17th week of amenorrhea, is recommended to the patient. The procedure, which consists in puncturing a sample of amniotic liquid during an ultrasound, is far from innocuous. The final result of the fetal karyotype is obtained about two weeks later. Besides, in nearly 3% of the cases, a second sampling is necessary because of difficulties of interpretation [7]. The iatrogenicity is so important that the amniocentesis results in the loss of two unharmed fetuses for the screening of one affected one. All in all, the iatrogenous risk of miscarriage, by no means insignificant, fluctuates between one and two percent. In France, 700 healthy fetuses, on average, are estimated to be lost annually because of this method.

Age	**Number of women tested**	**Number of women at risk**	**% at risk women / women tested**	**Number of Trisomy 21**	**% Trisomy 21 /at risk women**
< 38 years	637 783	43 655	6,8	352	0,8
=or>38 years	27 271	11 986	44,0	164	1,4
Total	665 054	55 641	8,4	514	0,9

Figure 10. Number of women tested with maternal serum markers and considered at risk at the threshold of 1/250 in 2007 in Agence de la biomédecine. *Rapport annuel et bilan des activités, diagnostic prénatal 2007*. p 280

Because of the rates of false positives, the number of amniocenteses is excessive: 95,452 in 2007. France holds the world record in percentage of pregnancies, which is 3.3 to 11.7%, even 16% in some medical services in the Paris area [7-8]. For several years, we have being using the Dr Brideron calculation Method of integrated risk, combining M.S.M., nuchal translucency, the cranio-caudal length of the embryo and the age of the patient. However, the calculation of an integrated risk has never been validated [9].

Aiming for increased efficiency, the French Minister of Health has recently modified the legal framework of PND and introduced a combined screening in the 1st term. The ministerial decree of June 23, 2009 suggests a change in the screening method for Trisomy 21 by combining the dosage of two serum markers (PAPP–A and free choriogonadotropin β subunit with hCG), the measurement of nuchal translucency (both subject to a quality control) and the maternal age [10]. For a detection rate of 85%, the rate of the false positives would be inferior to 5%. The new text defining the details of implementation of the new early screening for Trisomy 21 follows some

American recommendations as well as the results of a study in several centers in the Paris area. [8-11].

To conclude this first part, it is useful to note that a dosage of serum markers as well as an ultrasound are simply predictive or probability screenings. It is far from easy to announce a possible fetal abnormality when something wrong on the image or in the numbers doesn't a fortiori mean there is any abnormality. Is it ethical to worry a pregnant woman, parents-to-be with no positive proof? More than a century ago, Claude Bernard prophesied such an ambiguity of a medicine turned predictive: "Truth is the aim, plausibility is the risk" [12]. This is particularly evident in the screening for Trisomy 21. The current methods of identifying risks, based on probability, present couples with agonizing and unpredictable issues. As a consequence, couples resort massively to such invasive procedures as amniocenteses to confirm or validate the degree of uncertainty, with the collateral damages we are familiar with. The statistical modeling of intrauterine life is not just a theory in the Greek etymological meaning of the word which means contemplation. It has now become essential as an interface - as essential as the computer screen on which the doctor scrupulously enters the clinical data of his patient. We are looking away, focusing on numbers.... Are our medical views changing? How do we summarize the case history of a pregnant woman at the center meetings? Having first recalled the patient's history, the date of fertilization, the doctor then gives the numbers corresponding to nuchal translucency and statistical risk. Numbers which become the fetus' sole identity... without any legal identity!

Is statistical science on the point of replacing medical morals and of educating our consciences, since our actions - medical or iatrogenous terminations of pregnancy - based on calculations would consequently be legal? [13] What conscience of humanity does this statistical science bring us? What rules does life teach us which allow us to attempt to control it in this way? Is our recent medical approach to pregnancy irreversible then? The existence of the fetus based on numbers, a threshold value for maternal serum markers similar to a disqualifying mark (when the risk is superior or equal to 1/250) have insidiously transported us to a new culture of the existence which has become trivialized. Both the embryo and the fetus have indeed become objects of calculation and have stopped being subjects of care. Can it not be said that what medicine has gained in scientific knowledge, in statistical demonstration, in "chilly rigor", it has lost in the opportunity to "meet" the Other? When scientific obstetrics becomes more interested in the illness than in the ill patient, it may well lead us to forget the Other, the human person.

Does this mathematical specialization not threaten to make the health care professional less aware of the patient, be he or she a patient-to-be, and of the very meaning of care? Because they tend to turn the fetus into an object, might the evolving techniques of PND go against the very nature of care? One cannot end this chapter without quoting François Rabelais: "Science without conscience leads to the ruin of the soul!..."

PND – The Eugenics Issue

1. A Renewed Ethical Question

According to Hippocratic ethics, a doctor must be determined in his fight against disease. In such a merciless fight, he sometimes has to resort to drastic solutions. Modern medicine has intensively tried to suppress infectious diseases and eradicate the pathogenic agents which carry them. The reward for such efforts is for example the disappearance of the smallpox virus everywhere on the planet, surviving only in some highly specialized laboratories throughout the world. But is this radical logic, deemed necessary for infections, not wrongfully being imported into the field of human diseases affecting fetuses? An eradication of chromosomal abnormalities and of other genetic mutations before birth seems to have become the ultimate aim of modern biomedicine - except that in this case the *carriers* must be eliminated, raising the haunting issue of an eugenics peril. Even though it reminds us of dark times in our history, the issue of eugenics is making a spectacular comeback in France in current ethics and policy debates.

In accordance with the first bioethical law of July 29, 1994, article 16-4 of the Civil Code states that "all eugenic practice leading to the organization of the selection of persons is forbidden". Although the ban is clear, the current strategy of prenatal screening has for some months given rise to some renewed ethical concern about a modern recurrence of eugenics. Professor Didier Sicard, then chairman of the National Consultative Ethics Committee (NCEC), the most prestigious institutional group supervising the French moral debate, alarmed by the excesses of our public health policy about PND, has thus shaken our certainties : "Let us dare to state it: France is progressively building up a health policy that flirts with eugenics. (…) I am deeply worried by *systematic* screening. How can one defend the right to non-existence? (…) We are not far from the impasses we began to enter into at the end of the nineteenth century where science might come to decide who should live and who should not. And yet, history has amply shown where it can lead us when one starts to undertake the exclusion of groups of humans from the city, based on cultural, biological and ethical criteria. The central truth is that most of the prenatal screening activity aims at suppressing, not treating. That is why such screening leads to a terrifying prospect: that of eradication" [14].

Demonstrating that even the Head of the State is now overcome with doubt, on February 11, 2008, the French Prime Minister asked the State Council, the highest administrative court of law, for "a detailed examination of the activities of prenatal screening". In a letter of mission he handed the Council to prepare the review of the law on bioethics, the Head of the government goes so far as to question the Council on whether "the precautions taken around these practices guarantee an effective application of the principle banning any eugenic practice".

2. *Diagnosis* or Prenatal *Screening*?

It is important to make a clear difference between the notions of diagnosis and screening. To make a diagnosis is a medical and individual act. Screening is a political and collective act, often initiated by public authorities. Its original aim is not to make a diagnosis but to give information about a statistically-calculated risk. As a consequence, in screening, what matters is not the demand but the offer as it imposes *in fine* the obligation for results. If diagnosis is indeed a noble part of the medical art, screening is far from being always necessary. To justify its practice, ethics requires that it result in a therapeutic response and corresponding attitudes. There is absolutely no denying the invaluable contribution of scientific progress in the area of prenatal diagnoses. Our improved knowledge and interpretation of maternal serum markers are of undeniable assistance in confirming an increased risk of preterm birth or of preeclampsia. In the case of intrauterine growth retardation, genetic diseases in the family and so many others, PND must be done as early as possible. Thanks to efficient tests, the obstetrician is able to inform couples, prepare them psychologically, guide them and if need be, take care of the baby in the best possible conditions in the event that the handicap might be cured by surgery. However, do all diagnoses, whether confirmed or suspected, lead to any other solution that the MTP requested by parents?

3. Individual Screening or Public Health Policy?

In a report released on May 6, 2009, the State Council didn't dodge the unsettling question of the Prime Minister and for the first time admitted that eugenics might not only be "the fruit of a policy deliberately followed by a state" but also "the collective result of a number of individual convergent decisions made by the parents-to-be in a society where the quest for the perfect child would prevail "[15]. Indeed according to the Council, some statistics show the existence of "an individual practice of elimination that is almost systematic". They go on to quote numbers to confirm such an appalling record: "In France, 92 % of cases of Trisomy are detected against 70% on average for all of Europe, and 96% of the cases thus detected lead to a termination of pregnancy". How can we explain the almost absolute link between PND and the MTP that ensues if the fetus is believed to be affected by "a particularly serious affection"?

Professor Jacques Milliez talks of a diffuse culture in our society which refuses the very notion of handicap. "It is generally accepted that, except in cases where parents disagree because of their beliefs or feelings, a fetus affected with Trisomy 21, can, legitimately in the sense of collective and individual ethics, undergo a MTP. There is some general assent, collective approval, consensus of opinion, established order in favor of such a decision, so much so that the couples who will have to undergo a termination of pregnancy for Trisomy 21 will not be asking themselves the difficult question of how relevant their individual choice is. Society and in some way, general opinion, even without any constraint, have answered the question for them. Practically everybody would have done the same. The indication seems even so established that the parents consider that somehow it is a right…We will not ask here any nagging question about the relevance of their choice" [16].

If eugenics is today described as democratic and liberal because it would only be the result of a vast number of individual and private choices, is it not as well a political and collective choice with individual consequences? According to Dr Odile Montazeau and Dr Josée Benoit : "If the spring loading mechanism of eugenics is indeed individual, it is put in place by public authorities" [17]. They conclude with these unkind words : "A MTP for Trisomy 21 has really become a eugenic practice and with a very large social consensus, in spite of the lack of – or maybe thanks to the lack of – any democratic debate (…) A MTP has become a tool to select unborn children, (…) one of the eugenic practices produced by a policy that doesn't admit it is so and that pretends to respond to the demands of the couples." Indeed, many observers note that behind such practices lies some kind of voluntarism on the part of public authorities, aiming at profitability and technical efficiency.

4. PND as a Profitable Practice?

Our Public Health policy about PND and Trisomy 21 is implicitly based on a comparison of the costs of screening against the costs of life-long care for children with Down Syndrome. A study released in 1993 estimated that the financial cost for society to take care of a person with Down Syndrome all his/her life amounted at the time to 2,650,000 francs [18]. Compared with what children who would not have been detected would have cost - some speak of an escape rate - some people think that a screening policy unburdens public funds (the screening process costs the French health insurance 100 million euros a year) [19]. This is what the very serious High Committee of Public Health says: "The analysis of cost and profit, when it simply compares the collective cost of amniocenteses and karyotypes against the cost of taking care of handicapped children who wouldn't have been screened, (and assuming that a positive diagnosis is systematically followed by a MTP) shows that PND are perfectly justified for the community" [20]. Moreover, in a highly publicized legal case in France, the Perruche case, where the doctor was held responsible for the handicap of the child, a case of congenital German measles that had not been diagnosed, the French Social Insurance had also registered a complaint, asking for compensation for loss due to the cost of taking care of the child. According to Jean Marie Le Méné, President of the Jérôme Lejeune Foundation, state-approved for research on genetically-transmitted mental deficiencies, that is all very logical. In an essential book on the issue, he says that society has invested everything in screening, has already paid for handicapped children to disappear and is not then going to pay for them to live [21]. Besides, economic necessity is such that, despite the invasive risks of amniocentesis, the death of healthy fetuses is tolerated. Finally, though tests are not coercively imposed on pregnant women who still have to give their written consent, they are constraining for doctors who are obliged to recommend them. Is it not because of this fact that the whole system has become so extremely legalized, turning the practice of an obstetrician into some logic of guaranteed results? If a practitioner fails to detect a handicap in an unborn child, he/she is today liable to a penalty... up to ten years after the child comes of age!

The individual versus society: duality or synergy of their interests? According to Claude Bernard: "A doctor always deals with an individual. He is not, in any way, the doctor of mankind, of the human race; he is the doctor of an individual and even of an individual in particular circumstances… [22]". Conversely, one can foresee the dreadful consequences of a true « gynocide » when some Indian or Chinese doctors accept to practice sex-selection abortion. We must admit that obstetricians of the beginning of the twenty-first century are not exclusively in the service of individuals. Society has invested them with a very particular mission – to terminate some pregnancies in order to protect the community's interests. That used to be the task of the executioner when the death penalty existed... Are we not confusing roles?

Both *faber*, thanks to his tools and methods, and *economicus* , as determined by his productivity, has Homo Sapiens turned into *homo statisticus* since 1997, as he has broken through the filter of prenatal statistical calculations ?

5. PND: A Method Achieving Greater Results snd Practiced Earlier snd Earlier

As we pointed out earlier, the High Authority on Health recommended generalizing a new screening system, carrying out all analyses at once during the first trimester while combining the dosage of serum markers with the measurement of nuchal translucency. From the advantages put forward - a lower amniocentesis rate and, as a consequence, fewer secondary miscarriages and a lower psychological impact in case of a MPT - , the French government implemented these recommendations in the decree of June 23, 2009.

While the new procedure is supposed to refine the theoretical increased risk of Trisomy 21, its positive predictive value has not yet reached a hundred per cent. Once again, screening is merely a calculation of risks. If the risk is low, screening won't absolutely dismiss the possibility that the unborn child be affected. If the risk is believed high, the diagnosis will have to be confirmed by an invasive act in all cases. But an amniocentesis being too uncertain in the first trimester, the obstetrician will have to propose to women a choriocentesis, or biopsy of the trophoblast (future placenta), possible by the twelfth week of amenorrhea, that causes a rate of fetus loss at least equal to the one caused by a puncture of amniotic fluid. Good practice recommends that only experienced practitioners should undertake the sampling of chorionic villi because there is a risk of miscarriage which is by no means insignificant. Even if it is theoretically limited, there is still some iatrogenic risk. Can we really accept to lose even one fetus on the mere presumption of a statistical risk?

What about the psychological impact on women? Will the new procedure amount to less anxiety for the mothers? According to the new decree, pregnant women must have their say in the choice of a diagnostic method. Either they will choose a biopsy of the trophoblast, earlier and at least as iatrogenic as an amniocentesis – however may not a supposed precocity of samplings turn out to be illusory as sampling centers are bound to get saturated ? - or they will choose the latter test, later and so are likely to become more anxious as they will wait longer for a diagnosis. In addition, won't the act, becoming ordinary, be all the more readily accepted by the patients as they will not yet have registered their pregnancy or become affectively involved? [24]. Can we underestimate the psychological consequences of such an act, when Post Abortion Stress Syndrome is getting more and more attention ?

Moreover, all the tests being taken one after another might lead to some technocratic type of management of screening, all happening in the same place. Some speak of a eugenic time-space concept to describe this change of practice. In France, a pilot project of this kind, the “Prima Facie” Center, opened in Paris in April 2009. Following a screening method still called “one day test”, it proposes to any pregnant woman a general checkup, clinical, biological and ultrasound testing, between the 11th and the 14th week of amenorrhea, with results given within 3 hours [23-24]. One could view this as a concrete implementation of a surprising proposal made by a famous Nobel Prize of Medicine winner, Sir Francis Crick, who suggested that any intrauterine being should not be considered human before tests have been done for its genetic endowment. If the tests fail, it loses the right to live [25]. Isn’t “Prima Facie” the Latin for either first sight or first piece of evidence?

Furthermore, the new time limits allowed by the Ministry of Health fall within the legal period for a termination of pregnancy in France, allowed till the 14th week of amenorrhea by the law of July 4, 2001. Women could then get around the issue of an approval for a MPT by a Multidisciplinary Center, by individually asserting their “right” to a termination of pregnancy if there is any suspicion of fetal malformation, even though there be no certainty of that kind of thing. Foreseeing some underlying danger in instituting the procedure, the State Council has proposed to compel women with at risk pregnancies to wait for the enlightened opinion of the commission of a Multidisciplinary Center for PND. For all that, nobody will be able to compel them to submit to such a measure prior to the usual date for a registration of pregnancy.

Lastly, the State Council recommended giving more information to pregnant women on PND in order to limit the drift towards eugenics. Wishful thinking, one might say! How can we assert that a woman who in a matter of hours will be informed that there is a risk her child might be affected, will then be able to really consent to such or such a test? And is her consent always free and enlightened? Do we express ourselves in an intelligible language? By the decree of May 6, 1995, along with a PND, a doctor must as well give full information to the pregnant woman. The health professional must attest he/she informed her when screening was proposed (certificate signed by the consultant), and written consent of the woman for screening must be attached to the prescription. Yet a study by the Inserm 912 Unit, led by Valérie Seror, one of the best specialists on the subject, has shown that most women were not aware of the implications of screening. In particular "40% of them had not considered that at some time they could have to be confronted with the decision to terminate their pregnancy. More than half of them had never thought that screening could be followed by an amniocentesis and about one third didn't understand the results of blood dosage" [8-26]. According to Valérie Seror: "One may argue that women are not aware of the potential implications of screening". Some observers have rightly noticed that this disturbing study has been little commented upon by the concerned authorities, as it questions the principle of enlightened consent which is theoretically the cornerstone of the statutory procedure that controls the screening of Trisomy 21.

The new French regulations themselves could very quickly become obsolete due to the analysis of fetal DNA in maternal blood at a very early stage of pregnancy. A test called ISET -isolation by size of epithelial tumor cells- initially perfected in oncology, has been experimentally validated in the field of obstetrics and PND. The original aspect of the method lies in the fact that a few epithelial tumor cells can be isolated individually by filtration, then be micro dissected and their DNA extracted. The person in charge of the study, Professor Patricia Paterlini-Bréchot, head of the Inserm 807 Unit at the Necker Hospital for Sick Children, drew the following conclusion on January 27, 2009, after studying the test for the detection of mucoviscidosis : "Not only was the right diagnosis made (affected / non-affected) in every case, but the method has also distinguished the fetuses carrying the disease (a mutated allele and a normal allele) from those that are completely normal, which shows how extraordinarily accurate and specific it can be" [27]. The method could be applied to any genetically-transmitted disease or chromosomal abnormality. Above all, specialists are anticipating the same degree of success for the detection of Trisomy 21. A universal and non-invasive method that would systematically be proposed to any pregnant woman may thus become available. Pregnant women could soon be able to decide early pregnancy termination regardless of the opinion of a college of experts.

CONCLUSION

On the one hand, obstetricians have no intention of posing as censors or of expressing a value judgment on the couple deciding to terminate a pregnancy after being told of a fetal malformation or chromosomal abnormality. How could they really be sure of their own reaction if personally confronted with the same situation? What would their attitude be if they were suffering as much and feeling as distraught as the couple they are ethically bound to support?

On the other hand, obstetricians may rightly question the evolution and moral meaning of the current change of PND. Is it now in the service of a vitalist medicine or of a utilitarian medicine founded on a science of probability? They have the unquestionable feeling of having become

sorters of future beings and of taking part in a real hunt for handicaps, as is the case with the screening of Trisomy 21. Must they now consider themselves to be "health officers"? They have been allotted this part by society itself. Would they dare to refuse screening? Certainly not! They know all too well that the birth of a child with Down Syndrome, by the mere fact that screening was not proposed to the couple, can lead to civil action. Obstetricians find themselves caught in a stranglehold don't they? Their responsibility and not the politician's or the legislator's is indeed involved. It all works as if some conditions had to be met to give proof that the process of human life can persist. The burden of proof belongs to the unborn child and obstetricians stand at the crucial spot of the system favored by the authorities. The specialty is being given a new epistemological definition overruled by a biocracy or "biopower" according to Foucault's word.

Obstetricians are essential actors in intrauterine life as well as in the evolution of society. How can they then assume their responsibilities despite the contradictions pointed out by this chapter? Have they become the tool of a medicine that is dehumanized... *ab ovo*? "In twenty years, the value of human life has infinitely deteriorated... In a world where rights are being accumulated, the fetus and the new-born child are only granted the right to live to the extent that we concede it to them" [28]. Is not Professor Y. Malinas' analysis still both fully relevant and topical in 2010?

Acknowledgments

The authors sincerely wish to thank Mrs C. Gamois-Boucabeille and Mrs MM Simoni for the translation, Dr C. Eglin for the ultrasound iconography and Pr D. Querleu for revising this chapter.

References

[1] Leblanc P, Arduin PO. L'éthique médicale à l'épreuve de la loi de Bioéthique. Point de vue d'un obstétricien. *J Gynecol Obstet Biol reprod* 2009 ; 38, 363-366

[2] Code de la santé publique, article L. 2131-1.

[3] Agence de la biomédecine. Rapport annuel et bilan des activités, 2007, Centres pluridisciplinaires de diagnostic prénatal, p.281.

[4] Souka AP ; Von Kaisenberg CS ; Hyett JA ; Sonek JD ; Nicolaides KH. Increased nuchal translucy with normal karyotype. *Am J Obstet Gynecol* 2005; Apr 192(4), 1005-21.

[5] Arduin PO. *La Bioéthique et l'embryon.* Paris, Edition de l'Emmanuel, 2007.

[6] Sureau C. Il est déjà un homme. *Rev Prat Gynecol Obstet* 2008 ; 128 :7

[7] Agence de la biomédecine. Rapport annuel et bilan des activités, 2007, Centres pluridisciplinaires de diagnostic prénatal, p.277.

[8] Seror V, Ville Y. Prenatal screening for Down syndrome : women's involvement in decision-making and their attitudes to screening. *Prenat Diagn* 2009; 29:120-8.

[9] http://pro.gyneweb.fr/jmb/gyneweb-echo/aneuplo/RT21.html

[10] Arrêté du 23 juin 2009, JORF n°0152 du 3 juillet 2009, p. 11079 .

[11] Reddy UM, Mennuti M. Incorporating First-Trimester Down Syndrome Studies Into Prenatal Screening. Executive Summary of the National Institute of Child Health and Human Development Workshop. *Obstet Gynecol* 2006; 107: 167-73.

[12] Riese W. *La pensée morale en médecine.* Presses Universitaires de France. Paris ; 1954, p. 5.

[13] Leblanc P. Le diagnostic prénatal, un progrès médical *? Lett Gynecol* 2008 ; 331 :6-9.

[14] Sicard D. La France au risque de l'eugénisme. *Le Monde* 5 février 2007.

[15] Conseil d'Etat. La révision des lois de bioéthique. *La documentation française* 2009, p. 40.

[16] Milliez J. *L'euthanasie du fœtus, médecine ou eugénisme ?* Odile Jacob. Paris ; 1999.

[17] Montazeau O, Benoît J. A propos de la clause de conscience. *Les dossiers de l'obstétrique* 2008/11 ; 35, 376 :3-8.

[18] Seror V, Moatti JP, Müller F, Le Gales C, Boué A. Analyse coût-bénéfice du dépistage prénatal de la trisomie 21 par marqueurs sériques maternels. *Rev Epidemiol Santé Pub 1993* ; 41 : 3-15.

[19] Rapport public de la Cour des comptes, *La vie avec un handicap, 2003.*

[20] Haut Comité de la Santé Publique. *Pour un nouveau plan périnatalité* Editions ENSP (Paris) 1994.

[21] Le Méné JM. *La trisomie est une tragédie grecque.* Salvador, Paris, 2009.

[22] Riese W. *La pensée morale en médecine.* Presses Universitaires de France. Paris; 1954 p.57

[23] Ville Y. La gestion des risques obstétricaux au premier trimestre : bien plus qu'un dépistage. *Med Fœtale Echogr Gynecol* 2008 ; 76 : 59-63.

[24] Ville Y. Primae Facie : la précocité du dépistage prénatal impose son professionnalisme. *J Gynecol Obstet Biol Reprod* 2009 ; 38 : 5-8.

[25] Guillebaud J-C. *Le Principe d'Humanité.* Paris, éd du seuil, 2001 (coll. Points, P1027, p.332)

[26] Seror V. Le diagnostic de la trisomie 21 est-il bien compris par les femmes ? Rapport Inserm 912, Marseille, 2009.

[27] Collectif interassociatif autour de la naissance. *Diagnostic prénatal : validation du test ISET de la mucoviscidose.* Communiqué de presse, 27 janvier 2009.

[28] Editorial published in *La Presse Médicale* 1986 ; 8 : 3-4.

In: Bioethics: Issues and Dilemmas
Editor: Tyler N. Pace
ISBN: 978-1-61728-290-4

Chapter 3

THE 'REAL WORLD' OF ETHICAL DECISION-MAKING: INSIGHTS FROM RESEARCH

Pam McGrath*
Central Queensland University, Brisbane Queensland, Australia

In recent years the literature on bioethics has begun to pose the sociological challenge of how to explore organisational processes that facilitate a systemic response to ethical concerns. In this chapter, I seek to make a contribution to this important new direction in ethical research by presenting an overview of my program of research that explores notions that inform health professionals' ethical decision-making in the 'real world' of health care organisations. The insights gained through the research are translated into practical strategies that can be utilized to strengthen the process of ethical decision-making in the health care system.

The organisational context for the research reported in the chapter is varied and includes insights from studies in acute medicine, emergency medicine, Indigenous health, obstetrics and mental health care. The findings from research conducted over this diverse range of psycho-social health concerns indicates that a core belief informing health professionals' ethical

* Dr Pam McGrath B.Soc.Wk., MA., Ph D, NH&MRC Senior Research Fellow, Director, International Program of Psycho-Social Health Research (IPP-SHR), Central Queensland University, Brisbane Office: Level 2, 147 Coronation Drive, Milton 4064, Brisbane Queensland, Australia, Postal Address: PO Box 1307, Kenmore Qld 4069, Australia, Tel: (07) 3025 3377. Fax: (07) 3114 9461, E-mail: pam_mcgrath@bigpond.com, Website www.ipp-shr.cqu.edu.au

decision-making is the notion of patient-centred care – that is, the guiding principle that addresses their sense of the 'good' or the 'ought' is to act in a way that furthers the interest of patients and their families. This chapter explores a range of topics associated with how health professionals operationalise this principle in the context of service delivery including issues of multi-disciplinary team work, advocacy, resolving conflict and tension, obtaining informed consent, dealing with procedural distress, decision-making at end-of-life and the role of the ethics committee. The discussion affirms the importance of a psycho-social, rather than purely philosophical, perspective as a productive new direction in bioethical research.

The discussion in this chapter is situated in the recent paradigmatic shift in bioethics that acknowledges that the modus operandi for reflection of ethical issues has moved from the historical focus on a purely philosophical mode to a form of reasoning that embraces a wide range of knowledge and disciplines (Frank 2004; Nandi 2000; Rajput & Bekes 2002; McGrath 2006). Indeed, the International Association of Bioethics now defines Bioethics as the study of ethical, social, legal, philosophical and other related issues arising in health care and in the biological sciences (IAB, 2009; Singer, 1993). Such an approach breaks out of the limitations of restricting analysis to philosophical reflection (Livadas, 2002; McGrath, 1998a; 1998b) by broadening and enriching our understanding with the wealth of information and insights available from many academic disciplines. In particular, the chapter will provide an overview of a program of research in bioethics by the International Program of Psycho-Social Health Research (IPP-SHR) (www.ipp-shr.cqu.edu.au) that is providing leadership in demonstrating that the recent move to include the psycho-social perspective is producing major benefits for understanding the aetiology of, and formulating ways to address, significant health care ethical concerns.

The cornerstone of ethical reflection in Bioethics has traditionally been known as Principlism. This philosophical method involves the application of a set of four principles (autonomy, beneficence, non-maleficence and justice) to ethical decision-making (Beauchanp & Childress, 2001). While Principlism continues to provide a foundation for bioethical decision-making, there has in recent decades been an increasing critique of the exclusiveness and hegemony of that philosophically-based, predominantly abstract, rationalistic mode of reasoning in bioethics (Alderson, 1991; Bauman, 1993; Closer & Gert, 1990; DeGrazsia, 1992; Nicholson, 1994; Tong, 1996). The work of Carole Gilligan (1982) led the way to a new perspective in bioethical reasoning. Gilligan's work drew attention to the fact that the rationality of abstract philosophy leaves out important non-rational factors associated with the social context and relationships informing ethical decision-making. Her work placed notions of care and relationships firmly on the agenda as criteria for ethical reasoning contradicting their prior marginalisation in a system that exclusively valued abstract rationality. The spirit of this paradigm shift in ethical reasoning that underpins the discussion in this chapter is well summed up in the following statement by Johnstone (1994:100)

> A substantive moral theory or ethic is one that should be well grounded in the 'concrete circumstances of life' and not the 'detached fantasies of the hypothetical ideal'...Provocative questions are raised about the reliability of a value-neutral account of rationality, and why reason should be regarded as having any more authority in moral thinking than the moral sentiments of, for example, empathy, compassion, sympathy, friendliness or caring.

An outcome of this change in direction is that in recent years the literature on bioethics has begun to pose the psycho-social challenge of how to explore organisational processes that facilitate a systemic health care response to ethical concerns (Gallagher & Godstein, 2002). This approach is summed up by Cutcliffe and Ramcharan (2002) as the 'ethics-as-process' approach.

A Hospice Start

IPP-SHR's program in the psycho-social aspects of ethical decision-making began with a focus on the interface of acute and palliative care. This initial study explored the notion of autonomy in the context of end-of-life decision-making for patients and their families cared for by a leading Australian hospice, Karuna Hospice Service (KHS). The findings indicated that the major contribution of KHS is the discursive richness of its notion of autonomous choice which incorporates a holistic respect for the individual and the active creation of alternatives (McGrath, 1998a). KHS's ethical response is not through principles applied to difficult situations, but rather through a holistic interpretation of the provision of service where the ethical will is expressed throughout the organisational discourse.

Ethical praxis is directed towards empowering the client through the active provision of alternatives and a respect for the individual's right to handle dying in his/her own way. Support for different ways of responding to terminal illness (such as the KHS opportunity to die at home with appropriate palliative care rather than in an acute medical ward surrounded by escalating technology) is as important in seeking bioethical solutions as is the application of principles to the complex and inescapable ethical dilemmas after they arise in a technologically-driven, reductionist, medico-centric discourse of today's hi-tech response to dying. As one participant in the study summed up the notion of choice:

> In their own homes they are the boss, they are in control, not close to a lot of technology and not caught up in that hype. There is nothing like being in your own surroundings, being in your own bed. (McGrath,1993:525)

The creation of, rather than just reflection on, choices must be a consideration at the heart of bioethics, for as Veatch (1989: 3) explains, 'for people to make choices is at its core an exercise in ethics'. The findings are set in the context of the broad literature that addresses the powerlessness of the patient in the health care system.

From Hospice to High-Tech Care

The next step in the program was to turn the bioethical research focus on autonomous choice to high-tech health care; in this case, the informed consent issues in relation to peripheral blood stem cell transplantation (PBSCT) (McGrath, 2000). PBSCT, in which stem cells are mobilised in large numbers and harvested from circulation, is increasingly utilised as the transplantation modality in a range of both haematological and non-haematological cancers. The experience of transplantation can be a major stressor for both the patient and their family (McGrath, 2000). Transplantation involves painful medical procedures, anxiety about uncertainty of outcome, severe limitations for at least six to eight weeks and changes in body image and body functioning, as well as the risk of dying associated with the toxicity of high dose chemotherapy (McGrath, 2000b). Thus, consenting to PBSCT is a serious and life-altering decision not only for patients but also for their families. The ethical importance of ensuring a respectful and sensitive handling of the decision to participate in treatment through well defined informed consent procedures is essential.

The notion of informed consent is founded on the concept of respect for person and the principle of autonomy (Johnstone, 1994). Traditionally the key criteria informing the process of consent included full disclosure, comprehension, voluntariness, competence in decision-making and written confirmation of consent (Faden & Beauchamp, 1986). The procedure of obtaining consent was narrowly focused on the legalistic process of obtaining a signature on a consent form. This somewhat limited perspective has now increasingly been challenged by the psycho-social perspective that acknowledges the myriad of emotional and sociological factors that impact on the consent process.

The findings from our study clearly demonstrate that a myriad of physical, emotional, social and spiritual issues, rather than purely rational/philosophical considerations, impact on the patient's decision to give consent to undergo PBSCT (McGrath, 2000). At the point of decision-making patients are often grappling with the shock of receiving a life-threatening condition and struggling to accept the implications of such an illness. Information given to them about treatment will be filtered through the psychological processes of denial, fear, optimism and hope. The decision to undergo treatment is usually made in the hospital setting where, although patients feel great respect and appreciation for the hospital staff, they will also be vulnerable to feeling disempowered. The decision to undergo PBSCT for many is a *fait accompli* because of the stated lack of choice – the only alternative is perceived to be death. As most patients have experienced less demanding courses of chemotherapy prior to the discussions about PBSCT, their assessment of the side-effects may be underestimated. Participants in the study also report being unaware of, or unable to imagine, the difficulties involved in the long period of treatment and recovery. There were indications that some would not wish to, or agree to, undergo another PBSCT. Thus, the findings affirmed the notion of 'constrained voluntariness' (Jacoby, 1999) because the survival motive, trust in the physician and fear of worsening of the disease, rather than autonomy, are the most salient factors informing consent for PBSCT. The research highlights the importance of issues of power. The fact that patients do, to a large degree, see the decision to undergo PBSCT as a *fait accompli* places considerable responsibility on the doctor with regards to ensuring that the offer of treatment is in the patient's best interest.

As with the prior study on hospice care, the work on autonomous choice for PBSCT highlights a range of psycho-social and power issues that are at the core of the consent process. These issues undermine any notion of the efficacy of understanding the ethical process through abstract reasoning detached from the reality of the situation.

TURNING THE FOCUS TO MENTAL HEALTH

The next project that was initiated in IPP-SHR's program of research on ethical decision-making has a focus on informed consent issues during end-of-life care within the context of institutional mental health (McGrath & Forrester, 2006). In essence, the health workers at the institution felt constrained to implement resuscitation procedures with dying in-patients even when such actions were likely to cause considerable distress to the patient and staff and when there was not any likelihood of actually resuscitating the patient who was in the stages of dying. All of the participants in the study reported that the most significant factor impacting on their decision-making, as mental health care workers, in relation to end-of-life care was the prescriptive nature of the legal framework that defines their work. In particular, the necessities of a coronial inquest into all deaths at the institution and the legal imperative to engage in cardiopulmonary resuscitation (CPS) were the two factors that informed their response to patients during the dying trajectory. For the health care workers in this institutional setting, the legal framework of institutional mental health provides pressures and restrictions that inhibit the full integration of palliative care best practice end-of-life care. This was despite the fact that the institution's general policy and procedure document for care of terminally ill patients defines the outcome of terminal care in clear palliative care best-practice principles as: 'All deceased patients will be treated with respect, dignity and with due regard to their cultural and spiritual needs'. The results from the research indicate that the fears and restrictions imposed by the perception of the legal framework are the significant factors impacting on, and to a large degree inhibiting, the provision of best-practice palliative care in the mental health institution.

However, the health professionals' fears are in contradiction to the actual letter of the law. The Section 282A of the Queensland Criminal Code Act clearly states that a person will not be criminally responsible for providing palliative care if that care is:

a. provided in good faith with reasonable care and skill;
b. reasonable having regard to the person's state at the time and the circumstances of the case; and
c. provided by a doctor… or ordered by a doctor who confirms the order in writing.

Though the section expressly prohibits any action taken to intentionally bring about the death of a patient, it provides protection from criminal liability where the incidental effect of providing palliative care is to hasten the patient's death.

The findings highlight the quite significant obstacles that fear of and misconceptions about the legal imperatives associated with institutional mental health care impose on health care workers who wish to provide compassionate palliative care. Again, the factors driving ethical decision-making are a myriad of psycho-social and legal issues, rather than abstract philosophical reflection.

Patient-Centred Care: Acute Medicine Provides Insights for A New Direction

As can be seen by the summary overview, the research from the program up to this point showed that a myriad of factors other than rational reflection on a mantra of principles is driving ethical decision-making in the health care system. The question that then needs to be answered is: if ethical decision-making is not driven by the application of the quartet of principles known as Principlism, what modus operandi are health professionals using to resolve ethical problems in the system? The next IPP-SHR study began to explore this idea through research that focused on the ethical decision-making process of health professionals working in acute medicine. The study focused on the actual experience, understanding and attitudes of clinical professionals in acute medicine (McGrath, Henderson & Holewa, 2006). Findings from the study of relevance here pertain to how a multi-disciplinary team of health professionals defined and operationalised the notion of ethics in an acute ward hospital setting.

The key issue reported from the findings was that health professionals are not only able to clearly articulate notions of ethics, but that the notions expressed by a multi-disciplinary diversity of participants share a common definitional concept of ethics as patient-centred care (McGrath, Henderson & Holewa, 2006). The central finding is that for all professional groups there is a guiding notion to address their ethical sense of the 'good' or the 'ought', and that is to act in a way that furthered the interests of patients and their families. In short, the core definition of health care ethics from the perspective of the health professionals is that of providing patient-centred care. The findings affirm the importance of the psycho-social perspective as a productive new direction in bioethical research.

The findings from that study were that patient-centred care requires an ability to access the patient's perspective. Patient-centred care is based on consultation with the patient or, if necessary, family members acting on the patient's behalf. Consequently, effective communication is viewed as the medium for patient-centred care. There were a number of conscious strategies recorded by participants for achieving clarity of communication, including ensuring that information given is understood, and building quality relationships with the patient. Time and continuity were seen by nursing and rehabilitation staff as being an advantage and the lack of these factors seen as disadvantages for doctors, who have less time to spend with patients.

All professional groups perceived a multi-disciplinary team (MDT) approach as the core strategy to determining the wishes and circumstances of the patients. Such an approach embraces both formal and informal MDT relationships among health professionals within and outside the hospital. Case conferences and family meetings are seen as important forums for MDT communication.

To a minor degree, the language of the health professionals is informed by the academic ethical discourse of Principlism (Beauchanp & Childress, 2001), with participants making reference to the specific principles of autonomy, beneficence, non-maleficence and the derived principle of confidentiality. However, the findings indicate that the participants' discussion of ethical decision-making predominantly centred on psycho-social issues of communication, professional relationships and the multi-disciplinary organisational structure. Again, the findings affirm an 'ethics-as-process' view of moral reasoning (expressed as acting in accordance with the patient and family's wishes), rights and expectations in providing a high level of care. Interestingly, the findings indicated that, for health professionals, the link between patient-centred care and ethics is not yet overtly expressed in the ethical language of health care, but rather the assumption is made at a subconscious and not clearly articulated level. This research made *explicit* this *implicit* notion, as it is this idea that is actually driving the ethical decision-making at the coalface of health care. Thus, this study added to the growing body of evidence that the way forward with understanding and responding to ethical decision-making in the health care system is to affirm the importance of the psycho-social perspective, as opposed to purely rational philosophical reflection.

Dealing with Professional Conflict and Tension – An Ethical Issue

An important outcome of this new perspective on health professional's ethical decision-making is the understanding that although there is a shared conceptualization of ethics as 'patient-centered care', there can be times of conflict and tension in determining what is best for the patient (McGrath & Holewa, 2006). Thus, insights on the process of how a multi-disciplinary mix of health professionals respond to the ethical tension of such conflict and tension is at the core of understanding 'ethics-as-process'. In contemporary health care practice the MDT is seen as at the heart of patient-focused care (Drayton, Canter, & Allen, 2003).

An important caveat to the findings is that the group of health professionals involved in the IPP-SHR study (McGrath & Holewa, 2006) discussed in this section perceive conflict and tension as the exception rather than the rule. This insight was affirmed by the patients interviewed at the hospital who noted that the health professionals worked well together as a team. The findings indicate that at times when conflict does arise it is likely to involve tension at the interface of the nursing and medical professions.

The predominant concerns generating discord are end-of-life issues, especially regarding invasive technology. The nursing participants in the study reported that, as nurses are the professionals who spend long contact hours with the patients and who interface with the family, they consider themselves able to understand the client's perspective. The nursing concern is that, as the doctor spends only brief sessions with the patient, they are not always aware of the patient and family perspective and can at times not be prepared to listen to nursing recommendations. Such findings are affirmed by research by Coombs and Ersser (2004) that demonstrates that the nursing role, although pivotal to implementing clinical decisions, remains unacknowledged and devalued.

The findings from the IPP-SHR study (McGrath & Holewa, 2006) detailing issues associated with conflict and tension in a MDT recorded the wide range of ways that health professionals deal with such conflict and tension. One such strategy is the 'standoff' which is a breakdown in communication in which professionals in a dispute engage in dysfunctional confrontational silence that does not resolve the problem and tends to affirm the medico-centrism inherent in the status quo. Another strategy is to ignore the conflict, and the findings indicate that many issues do not reach the conflict stage as professionals choose to ignore potential disagreements to maintain the good will of collegial relationships. The ethical problem with such a strategy is that it prioritizes collegial harmony over patient and family advocacy.

A functional and effective strategy detailed by the health professionals is effective communication at several organisational levels as an imperative for resolving a dispute once conflict is generated. In relation to patient care, the communication should be with the patient and his or her family to clarify the client's perspective, either through direct consultation or through a family meeting. At a collegial level, health professionals reported turning for support and guidance to other members of their disciplinary team. For nurses such collegial consultation is essential, initiated from the start of a concern, and ongoing. The reasons for consulting are reflective discussions, support, debriefing, assessing group consensus and affirming a position. In contrast, doctors report using collegial consultation after the event to affirm the rightness of a decision. Allied health professionals also rely on collegial consultation, with the motivation for them being to access a sounding board for ideas, to test opinions and to debrief.

All groups noted that in very difficult cases they will go up a level and seek communication with someone higher up the hospital hierarchy. Allied health professionals noted the importance of going up a level but do this through the nursing hierarchy and rely on good relationships with the nursing staff to facilitate this.

The medical participants indicated that they prefer to handle issues at their own level but will consult up a level if there is patient resistance to medical recommendations. Doctors will only go up a level in the medical hierarchy, and only in extreme cases, such as if legal action is likely. One of the doctor's hope is that higher consultation will convince the patient of the rightness of the doctor's recommendation. Also, the medial expectation for going up a level is to access the advice of more experienced doctors, gain reassurance and ensures faster decision-making.

In contrast, nurse's discussions about going up a level were couched in the language of advocacy. Nurses will go up both the nursing and medical hierarchy. The rationale is to shift responsibility for action and to gain information to relay to the patient and family. Thus, although all professional groups made reference to the MDT as the *modus operandi* of patient-centred care, it was mainly nurses that used the language of advocacy as ethical praxis (McGrath, Holewa & McGrath, 2006). Advocacy is seen as core work for nurses and described in holistic terms as the representation of the patient in the context of the family through the processes of respecting choice, information giving, explaining, communicating, maintaining professionalism and providing comfort. The reason given by the nurses for the need for advocacy is that doctors had insufficient time to actively listen to the fullness of the particular patient's circumstances and made doctor-imposed decisions that did not take into account the patient's personal circumstances. The findings indicate that nursing advocacy through the process of representation of the patient and family's interest to the medical profession and other decision-makers higher up the hospital health care hierarchy is seen as the appropriate response to such a hegemonic situation. It is reported by nurses that such advocacy requires confidence, a sense of being 'right' about an issue, experience in dealing with other health professionals, sound practice knowledge and an ability to communicate. Professional and clinical confidence and experience are noted as necessary to successfully engage in the process of advocacy. The findings indicate that the adoption of an advocacy role by nurses expressed their ethical praxis of patient-centred care and thus represents an important means through which MDT ethical decision-making can be enhanced, medico-centrism limited and patient-centred care improved. The outcome of successful nursing advocacy is seen to be consumer satisfaction with their health care.

The consumers in the study affirmed the professional insights by noting the doctors' lack of time to talk and listen which did at times lead to the imposition of medical decision-making and lack of involvement of the carer in discussions. However, the consumer also noted that the ability to communicate varied for nurses as well as doctors and was seen to depend on personality characteristics. All of this comment was set in the high level of satisfaction with the care provided by the hospital.

A recommendation from the study is that to foster ethical decision-making based on a patient-centred focus it is important ensure that the MDT has a democratic, flat structure where the team works together with collegial respect and effective communication. Such a recommendation affirms Borthwick and Dowd's (2004) call for new ways of inter-professional working that focus on multi-professional collaboration and challenge the present medical dominance. Most significantly, however, the findings affirm Irvine and associates' (2004) recommendation that it is essential to conceptualise ethics as a shared social practice, with a dialogic approach to ethical decision-making that places greater emphasis on open deliberation and the articulation, negotiation, exploration and generation of new ethical perspectives in the here and now of clinical practice.

The hope and expectation is that, by keeping the research focus on what health professionals in a MDT actually do, rather than on how they philosophically think about ethical dilemmas, the insights can be translated into practical strategies that can be utilized to strengthen the process of ethical decision-making within the health care system.

End-of-Life Issues – The Most Challenging Ethical Issues for Acute Medicine

Further findings from the study (McGrath & Henderson, 2008) indicate that all professional groups in this study of an acute medical ward find end-of-life issues the most challenging of all the ethical concerns. The study highlights the important sociological factors that impact on the ethical decision-making of acute care professionals in relation to end-of-life care and how understanding these factors can illuminate many important palliative care practice issues. The findings reversed the assumption that to understand palliative care ethical issues it is best to focus on the hospice and palliative care system, showing that significant palliative care issues can only be understood if the focus is on the acute care system. In short, the language of ethical analysis is sociological rather than purely abstract philosophical.

At the core of the ethical concern is the task of balancing the competing moral demands of the duty to treat versus the moral imperative to alleviate suffering. Participants in the study noted that this is a poignant issue on the acute medicine ward because of the high number of geriatric patients for whom medical interventions have limited efficacy and may have considerable negative impact on quality of life. The medical professionals were seen by the nurses to be inadequately trained in palliative care philosophy and practice. Doctors in the study referred to the lack of preparation provided during medical education with regard to dealing with end-of-life ethical issues. This is concerning, given that allied health professionals spoke of the importance of quality of life issues and the importance of an integration of palliative care. The important point is that it is the health professionals in the acute care system who are the 'gate-keepers' to palliative care. If they lack expertise and knowledge of palliative care then referrals are not made and a myriad of ethical problems arise in the acute system associated with the pressure to treat and the importance of respecting quality of life. This work is but another example of the importance of understanding the psycho-social dimension of ethical concern, not only for the resolution of ethical dilemmas but for strategies (in this case, effective palliative care referrals) that pro-actively deflect ethical problems before they arise.

Similar Issues in an Emergency Department

The research described above was initially carried out in the acute medical ward and then repeated in the Emergency Department (ED) of the same hospital to see if health professionals from different sub-cultures communicated different insights. The important overall finding from the study set in the ED is that it echoes the results from the earlier work set in an Acute Medical ward and thus provides further evidence to affirm the insights gained from this new direction in understanding organizational ethical processes (McGrath & Henderson, 2008). At the core of both sets of findings is the idea that the core dimension of the notion of ethics for the health professionals interviewed is the service delivery notion of providing the *best possible care for patients*.

In the ED, where the decision-making is described as medico-centric, advocacy *ipso facto* necessitates a challenge to doctor decision-making. The findings (McGrath & Phillips, 2009) indicate that ED nurses' experience with advocacy varied depending on the democratic qualities and communication skills of the particular doctor who had care of the consumer. It was noted that seeing the need for advocacy does not necessarily translate into effective action, as management support is essential for productive advocacy. A phenomenon of the desire not to rock the boat was reported. The findings indicate that the support of other nurses is essential for advocacy and affirm the importance of focusing on the ethical nature of the organisation as opposed to an exclusive focus on the individual.

The work on the ED affirms that the psycho-social perspective has been found to provide important and useful insights in understanding a wide range of ethical issues.

Maintaining the Momentum

Once defined, IPP-SHR's lens of psycho-social research continues to be applied to the many and varied issues that are ethically problematic in health care. To conclude this chapter, the discussion will focus on two quite disparate issues, Indigenous health and obstetrics, to demonstrate the richness, variability and applicability of the psycho-social approach to ethical decision-making.

Indigenous Informed Consent

Despite the extensive consideration the notion of informed consent has heralded in recent decades, the unique considerations pertaining to the giving of informed consent by and on behalf of Indigenous Australians have not been comprehensively explored. This deficit is concerning, given that a fundamental premise of the doctrine of informed consent is that of individual autonomy which, while privileged as a core value of non-Indigenous Australian culture, is displaced in Indigenous cultures by the honouring of the family unit and community group, rather than the individual, as being at the core of important decision-making processes relating to the person. IPP-SHR psycho-social research (McGrath & Phillips, 2008), situated in the context of the literature on cultural safety, highlights the difference between the Aboriginal and biomedical perspectives on informed consent.

The first difference documented is between the Western focus on the individual giving consent and the Indigenous cultural imperative that consent is obtained from the 'right' person within the community. This is an important difference in that Indigenous peoples are not culturally empowered to provide individual consent, as in the west, but must seek approval from appropriate people in the extended family relationships or from authority figures, such as the *Jungai,* in the community. Disrespect for such traditional processes can lead to payback for the ill Indigenous person or their relatives, as well as considerable anger from family members and clinical staff.

A second concern with regards to consent issues for Indigenous peoples is that there are strong cultural differences towards time. Western medicine is based on a sense of immediacy where the clinical condition dictates the speed of response. Thus, clinicians can see it as a waste of time to have to engage in a process of community consent for procedures in circumstances where it can take weeks or even months.

Thirdly, the Indigenous consent procedures can pose practical problems such as having to travel long distances to speak to the 'right' person. Forward planning, wherever possible, is seen as a key strategy to avoid problems of delay.

The fourth concern relates to the likelihood of Indigenous peoples consenting to medical intervention in circumstances where they have no real understanding of or experience with the associated consequences. Examples were given of chemotherapy procedures where Indigenous peoples can feel angry or terminate the procedure when they find the medicine makes them feel worse than the disease.

The fifth concern relates to the language barriers, pointing to the major problems with regards to obtaining reliable informed consent when Indigenous peoples are not fully informed in their own language about the benefits and harms of the intervention they are offered. The findings indicate that it is important to ensure that interpreters are available who can effectively translate medical information and who are trusted and have the confidence of Indigenous peoples.

The doctrine of informed consent has as its moral and legal underpinning the westernized notions of individual autonomy and self-determination. Such notions are consistent with the general beliefs underpinning modern, westernized culture but are contrary to the traditional values held by Aboriginal cultures, which emphasize the importance of familial and cultural groups over individual autonomy. Given that it is equally important to obtain informed consent to medical treatment from Indigenous and non-Indigenous patients, it is imperative that core measures be implemented to ensure that Aboriginal cultural values are respectfully upheld in the context of the provision of westernized health care. This is a clear example that ethical reflection cannot be based on purely abstract philosophical reflection but must embrace a wide range of psycho-social concerns, in this case that of cultural understanding.

Ethics in Obstetrics – Informed Consent to Birth Choice

The final example to be discussed in this chapter is IPP-SHR's research which explored, from the mothers' perspective, the process of ethical decision-making about mode of delivery for a subsequent birth after a previous Caesarean Section (McGrath & Phillips, 2009). In contradiction to the clinical literature, the majority of mothers in this study were strongly of the opinion that a vaginal birth after caesarean (VBAC) posed a higher risk than an elective caesarean (EC). From the mothers' perspective, risk discussions were primarily valuable for gaining support for their pre-determined choice, rather than obtaining information. The mothers' decision-making was not informed by clinical considerations, indeed the mothers were poorly informed of the clinical/biological implications of their choice. The findings posit ethical concerns with regards to informed consent and professional obstetric practice at a time when there is a documented and worrying trend towards an increase in births by caesarean section (CS).

The doctrine of informed consent is underpinned by the requirement that the individual has sufficient information about the options available so as to enable a rational choice to be made. The key requirements for the giving of informed consent are full disclosure of all relevant information; comprehension and understanding of that information; freedom or voluntariness; and competence in decision-making, with the co-existence of all requirements recorded by a written expression of consent (McGrath 2000; Lovat & Mitchell 1991). Clearly, for this group of mothers who had all provided written consent for their CS operations, prerequisites to the giving of informed consent were not met. Importantly, issues of relevant information, comprehension and understanding of that information were not prioritised. The ethical imperative in relation to autonomy for this group of women was support for their pre-determined choice. The important point is that psycho-social knowledge of the dynamics of decision-making casts a very different light on the factors contributing to consent to birthing choice. Thus, it is strongly argued that the way forward in understanding and resolving the ethical concerns about rising CS rates lies in further investment in psycho-social research to deepen our collective understanding of the factors actually contributing to this health care dilemma. Clinical knowledge and abstract reflection make a contribution but are only part of the solution.

Conclusion

This chapter has provided a wealth of research on a wide range of topics associated with ethical decision-making in health care that demonstrates the importance of the psycho-social approach to understanding such issues. It is the hope and expectation in sharing this work that it will contribute to maintaining the momentum for this innovative and productive approach to bioethics.

References

Alderson, P. (1991). Abstract bioethics ignores human emotions. *Bulletin of Medical Ethics*, 68, 13-21.

Bauman, Z. (1993). *Postmodern ethics*. Oxford: Blackwell.

Beauchamp, T., & Childress, J. (2001). *Principles of biomedical ethics*. New York, NY: Oxford University Press.

Borthwick, A., & Dowd, O. (2004). Medical dominance or collaborative partnership? Orthopaedic views of podiatric surgery. *British Journal of Podiatry*, 7, 2, 36-42.

Closer, K. & Gert, B. (1990). A critique of Principlism. *Journal of Medicine and Philosophy*, 15, 219-236.

Coombs, M., & Ersser, S. (2004). Medical hegemony in decision-making- a barrier to interdisciplinary working in intensive care? *Journal of Advanced Nursing*, 46, 245-252.

Cutcliffe, J., & Ramcharan, P. (2002). Levelling the playing field? Exploring the merits of the ethics-as-process approach for judging qualitative research proposals. *Qualitative Health Research*, 12,7, 1000-1010.

De Grazia, D. (1993). Moving forward in bioethical theory: Theories, cases and specified principlism. *The Journal of Medicine and Philosophy*, 17, 511-539.

Drayton, S., Canter, A., & Allen, C. (2003). Dilemmas in providing patient-focused care. *CANNT Journal*, 13, 4, 30-33.

Faden, R., & Beauchamp, T. (1986). *A history and theory of informed consent.* Oxford University Press, New York.

Gallagher, J., & Goodstein, J. (2002). Fulfilling institutional responsibilities inhealth care: Organisational ethics and the role of mission discernment. *Business Ethics Quarterly*, 12, 4, 433-450.

Gilligan, C. (1983). *In a different voice: Psychological theory and women's development*. Cambridge, MA: Harvard University Press.

Internatinal Association of Bioethics. (IAB). *http://bioethics-international.org/iab-2.0/index.php?show=objectives* Accessed on 20th November 2009.

Irvine, R., Kerridge, I., & McPhee, J. (2004). Towards a dialogical ethics of inter-professionalism. *Journal of Postgraduate Medicine,* 50, 278-280.

Jacoby, L. (1999). The basis of informed consent for BMT patients. *Bone Marrow Transplant*, 23, 711-717.

Johnstone, M-J. (1994). *Bioethics: A nursing perspective.* 2nd Ed. Marrickville: Harcourt Brace Jovanovich.

Lovat, T., & Mitchell, K. (1991). *Bioethics for medical and health professionals*. New South Wales: Social Science Press, New South Wales.

Nicholson, R. (1994). Limitations of the four principles. In R. Gillon (Ed.) *Principles of health care ethics* (pp. 267-275). Chichester: Wiley.

Frank, A. (2004). Ethics as a process and practice. *Internal Medicine Journal*, 34, 355-357.

McGrath, P. (1998a). Autonomy, discourse and power: A postmodern reflection on rationalisty in bioethics. *The Journal of Medicine and Philosophy*, 23, 5, 516-532.

McGrath, P. (1998b). *A Question of Choice: Bioethical reflections on a spiritual response to the technological imperative*. Hampshire, UK: Ashgate Publishing Limited.

McGrath, P. (2000a). Informed consent to peripheral blood stem cell transplantation. *Cancer Strategy*, 2, 44-50.

McGrath, P. (2000b). *Confronting Icarus: A psycho-social perspective on haematological malignancies.* Ashgate Publishers: Hampshire, England.

McGrath, P. (2006). Patient-Centred Care: Qualitative Findings on Health Professionals' Understanding of Ethics in Acute Medicine. *Journal of Bioethical Inquiry*, 3, 149-160.

McGrath, P., & Forrester, K. (2006). Ethico-legal issues in relation to end-of-life care and institutional mental health. *Australian Health Review*, 30, 3, 286-297.

McGrath, P., & Henderson, D. (2008). Resolving end-of-life ethical concerns: Important palliative care practice development issues for acute medicine in Australia. *American Journal of Hospice and Palliative Medicine*, 25, 3, 215-222.

McGrath, P., & Holewa, H. (2006). Ethical Decision Making in an Acute Medical Ward: Australian Findings on Dealing with Conflict and Tension. *Ethics and Behavior*, 16, 3, 233-252.

McGrath, P., Henderson, D., & Holewa, H. (2006). Patient-Centred Care: Qualitative findings on health professional's understanding of ethics in Acute Medicine. *Bioethical Inquiry*, 3, 149-160.

McGrath, P., Holewa, H., & McGrath, Z. (2006). Nursing advocacy in an Australian multidisciplinary context: findings on medico-centrism. *Scandinavian Journal of Caring Science*, 20, 394-402.

McGrath, P., & Henderson, D. (2008). 'Oh, that's a really hard question': Australian findings on ethical reflection in an Accident and Emergency ward. *HEC Forum*, 20, 4, 357-373.

McGrath, P., & Phillips, E. (2009). Ethical Decision-making in an Emergency Department: Findings on Nursing Advocacy. *Monash Bioethics Review*, 28, 2, 15.1-15.16.

McGrath, P., & Phillips, E. (2008). Western notions of informed consent and Indigenous cultures: Australian findings at the interface. *Bioethical Inquiry*, 5, 21-31.

McGrath, P., & Phillips, E. (2009). Bioethics and Birth: Insights on risk-decision-making for an elective caesarean after a prior caesarean delivery. *Monash Bioethics Review*, In Press.

Nandi, P. (2000). Ethical aspects of clinical practice. *Archives of Surgery*, 135, 1, 22-25.

Rajput, V., & Bekes, C. (2002). Ethical issues in hospital medicine. *The Medical Clinics of North America,* 86, 4, 869-886.

Singer, P. (1993). The international association of bioethics. *Medical Journal of Australia,* 158, 298-299.

Tong, R. (1996). Feminist approaches to bioethics. In S Wolf (Ed), *Feminism and bioethics: Beyond reproduction* (pp. 67-94). Oxford: Oxford University Press.

Veatch, R. (ed.) (1989). *Medical Ethics*. Boston: Jones and Bartlett.

In: Bioethics: Issues and Dilemmas
Editor: Tyler N. Pace
ISBN: 978-1-61728-290-4

Chapter 4

THE RIGHTS OF FUTURE GENERATIONS IN ENVIRONMENTAL ETHICS

*Carlo Petrini**
Italian National Institute of Health
[Istituto Superiore di Sanità], Roma, Italy

ABSTRACT

Today it is widely recognised that much of the harm done to the environment is not limited by geographical, political or time barriers.

From the point of view of environmental ethics, our concern for future generations arises mainly from their position of disadvantage: preceding generations can limit the opportunities bequeathed to later generations by causing irreversible harm and depleting resources.

This paper first introduces some historical references; the most significant problems posed by the rights of future generations in regard to the environment are then identified, the notions of "common heritage" and "capital" as referred to the environment are discussed and the relations between "sustainable development" and the rights of future generations are presented. Some social aspects, legal implications and the contributions of philosophers and experts on the question of intergenerational justice in environmental terms are then analysed. An attempt is finally made to define the most significant problems involved.

---

*Head of the Bioethics Unit, Office of the President, Italian National Institute of Health [Istituto Superiore di Sanità], Via Giano della Bella 34, I-00162 Roma, Italy, E-mail: carlo.petrini@iss.it, Tel. ++ 390649904021, Fax ++ 390649904303

Historical Notes

Origins and Development

Concern for the rights of future generations in environmental terms has evolved gradually (Butera 2001). Some mention of such rights can be found in Francis Bacon's "Novum organum" (1620) and in the work of Descartes. The utilitarian current of thought initially proposed by Jeremy Bentham and John Stuart Mill was instrumental in focusing attention on the rights of future generations and, consequently, the duties of present generations towards posterity, though only to the extent that utilitarianism includes in the assessment of overall benefit not only present generations but also those still to come.

However, although philosophers, politicians and jurists (Roussopoulos 2004) have for many years addressed environmental issues, it is only in the last few decades that the formerly widespread tendency to consider the environment as of purely technical-scientific interest has been overcome. Sociology turned its attention to environmental problems much later than it did to other issues (Buttel 1986) and began doing so only indirectly by analysing the emergence of environmental movements and how the relevant challenges were perceived by various social subjects, as though social phenomena can be explained only through social events (in accordance with Emile Durkheim's well-known perspective) (Durkheim 1895).

So far as the question of future generations' rights are concerned, it was only in the last quarter of the twentieth century that this issue was systematically examined, mainly in opposition to policies whose only aim was perceived as being the pursuit of seemingly unstoppable development (De Marchi, Pelizzoni, and Ungaro 2001, pp. 11-36). This process was fuelled by debates on demographic growth, on the exhaustion of resources (primarily energy, but also biological, genetic and natural in the broader meaning) and on the various forms of pollution (Dower 1998, pp.767-79; Grima 2004, pp. 546-50).

Concern for this question is therefore increasing, both because technology augments the incisiveness of our possible actions and because there is growing awareness of the connections between the different components of human activities and the environment (European Commission 2004).

Hans Jonas

If we look at the tools traditionally used in the history of ethics, the duty to respect the rights of future generations can be defended using both deontological (formal correctness and respect for universal principles) and utilitarian (the greatest benefit for the largest number of people] arguments.

Another strong argument that can be used when weighing the rights of future generations is responsibility.

Hans Jonas is unanimously regarded as the writer who contributed most to examining the notion of responsibility in the debate on bioethics. (Jonas 1979) His contribution was decisive in developing the emerging discipline of bioethics at the start of the 1970's, at a time when fears concerning threats to the environment and to mankind were particularly acute (Potter 1970; Potter 1971) and some, including Jonas himself, prefigured humanity's demise.

Underlying Jonas' analysis is the recognition that scientific and technological progress gives us an ever increasing potential to intervene. Jonas noted that the consequences of this potential may not always be predictable: he emphasised that the scale of the effects of human activities may be such that they run out of control, with devastating results. Jonas argued that the new and increasingly incisive potential of human activities calls for a new approach to traditional ethical principles: the major pillars of rights and freedoms enshrined in traditional ethics should no longer be associated with a prediction of hope, but with the fear of an impending threat, in other words the worst-case scenario. The result is a "heuristic of fear" that calls for responsibility. The Kantian imperative is thus reformulated by Jonas in the following terms: "Act so that the effects of your actions are compatible with the permanence of genuine human life" (Kant 1993).

Jonas' ethical views were nonetheless attacked in some quarters.

One early criticism was that an ethic founded on the natural world, such as that proposed by Jonas, could be countered, in a Kantian perspective, by objecting that man is a moral being not because he is alive but because he is free and rational. In other words, bringing everything back to nature carries the risk of keeping us in a state of animality, rather than humanity (Goffi 2001, pp. 818-19). This might in turn be countered by pointing out that Jonas constantly associated comments on striving for the preservation of life with considerations on striving for the absolute, which is a strictly human characteristic.

Another criticism levelled at Jonas concerns his proposal for a drastic reduction in technological development. A primarily "defensive" ethic of this kind could harm the principle of beneficence, since technology brings advantages that would be precluded by the implementation of indiscriminate embargos (Goffi 2001, p. 818-819). This can perhaps be countered by pointing out that the scenario that alarmed Jonas does not make him demonise progress: instead, he shows how the exercise of responsibility can be fostered thanks in part to the possibilities made available precisely by science and technology.

The Two Principal Problems

The well-being of future generations can be expressed both as rights *in personam* and as rights *in rem*.

The Problem of Identity

The question of rights *in personam* runs into difficulties owing to the fact that, in this particular case, the rights cannot be assigned to specific persons. Obviously, future generations (with the exception of our immediate descendants) do not exist at the time when current generations are concerned with them. For this reason some writers object to the expression "rights of future generations" and prefer the terms "responsibility", "duties" and "obligations". Legal tradition holds that "rights" are granted to living subjects who may exercise them. In the case of non-living individuals it is difficult to apply even the particularly topical considerations regarding the rights of the embryo: even those who do not recognise an embryo as a human being cannot deny its existence.

Future generations thus fall into a very particular legal category. The formula favoured by most writers is therefore to refer to our responsibility, our care and our obligations towards future generations rather than to their "rights".

The Problemof Temporal Distance

The question of intergenerational equity involves what Bryan Norton called the "problem of distance" (Norton 1982). This is concerned less with the "translocal" dimension than with the "transtemporal" aspects. International treaties and conventions can relatively easily address the first (geographical) aspect. The temporal aspect, which is related to intergenerational equity, instead poses untold challenges from the legal point of view. The parties involved do not yet exist, they are potential and anonymous individuals who, in abstract terms, make up the "future generations". Clearly, the granting of rights to such persons poses practical problems and calls for careful theoretical analysis.

Faced with the uncertainties of dealing with problems that are temporally remote, there are those who maintain that duties towards the future extend only to our immediate successors. This was the conclusion reached, for example, by John Passmore, based on the premise that it is not possible to foresee what damage may be caused to individuals in the distant future (Passmore 1974). This argument is, naturally, not shared by those who adopt a decidedly "environmentalist" stance, who point out that practices that affect the climate, for example, may have a negligible impact on the next generation but very weighty consequences for those further down the line.

"Common Heritage" and "Capital"

General Aspects

The expression "common heritage" is often used when discussing intergenerational justice as it applies to the environment.

The word "heritage" immediately conjures the transmission of goods between succeeding generations of a family, and thus historical continuity. This becomes interesting when the problem of intergenerational equity is addressed. The concept of "common heritage" has economic and legal origins that go back to Roman law (Thomas 1988 quoted in Ost 1995), and was also adopted by sociology, where the expression "cultural heritage" is used in particular. The expression "common heritage" entered the debate on protecting the environment after 1982, the year in which the Convention on the Law of the Sea was drawn up (UN 1982). It had already been used in earlier conventions and treaties on the conservation of natural resources, but in such general terms as not to have found its way into any legal regime, anchored as these tend to be to a logic of "property" and "national sovereignty" ill suited to the global nature of environmental goods (Humbert 1992, p. 187-96). Some regulations do nonetheless contain an explicit reference to the notion of "common heritage" with reference to the environment. In France, for example, the law of 7 January 1983 states that French territory is the common heritage of the nation (Asemblée Nationale de la République Française 1983)

Other documents promulgated in the 80's preferred instead to adopt the less precise "common concern of mankind" (Birinie, Boyle and Redgwell 1992, p. 120). However, both the expressions "heritage" and "concern" refer to a duty of care incumbent on all states, regardless of the occurrence of harm: to use an expression coined by François Ost, we are caretakers of "a heritage that passes through the present" (Ost 1995).

The expression is thus often somewhat vague and remains a generic concept, of the kind that might be used of a "good family man" wisely administering his descendants' inheritance. This vagueness stems in part from the difficulties that arise when seeking a definition of the "common heritage" in positive law. Some of the concepts involved, such as the temporal distance already mentioned, are difficult to transpose into a legal context.

From the ethical point of view, the notion of "heritage" seems to belong to an extreme anthropocentric approach. In fact it conflicts with both extreme anthropocentrism and its opposite, deep ecology. The concept of environmental protection generated by this notion is of an "interactive" type; in other words, it acknowledges man's responsibility to the environment as a good to be safeguarded, respected and enhanced in the awareness that mankind is himself a part of his environment.

Because it is a generic concept, the notion of "common heritage" cannot easily be applied to specific cases, but can still be helpful when interpreting problems and seeking solutions.

It can thus be said that while the notion of "common heritage" currently has no precise legal features, it can certainly play an important role in environmental law. It may also help to bridge the gap between individual and general interests and between present interests and those of future generations. In this sense, the "common heritage" carries a lot of weight for the principle of equity.

Subject and Object

The term "common heritage" refers to a dualism between subject and object. At times the subjective aspect is emphasised, in other words the owner of the heritage and the means of transmitting it. This is, for instance, the predominant approach in French law. Elsewhere the objective aspect is stressed, with emphasis on the object itself to underline the notion of "common". The latter view prevails, for example in German environmental law (Ost 1995).

Whether the accent is on "filiation" or on the "common good", the derivation is a more general criterion of "good use" or management, similar to that of the prudent paterfamilias who does not waste, impoverish or make risky investments.

The two approaches, "heritage" and "common", have different connotations. Filiation is exclusive, and follows a line of descent safeguarded by legal regulations that ensure its protection from undue access by other subjects.

The "common" heritage, on the other hand, is not exclusive but inclusive: it is not a right "of" or "over", but a right "to".

The management of an exclusive heritage is handled differently from that of a common good. In the former case the aim is generally to maintain it, while in the latter case the aim is to ensure fair access by all. We can justly assert that the two approaches are blended in the management of environmental resources.

"Capital" snd "Human Capital"

The expression "common heritage" is sometimes associated with "human capital". In recent years it has been used in various institutions, occurring often, for example, in European Union projects.

The notion of "human capital" is a broad one. In his essay "Human well-being and the natural environment" (Dasgupta 2004) the Indian-born economist Partha Dasgupta emphasises how mere economic indicators are not enough to describe the level of a country's development. In his view human and natural elements should also be given broad consideration, as together they define a measure of "sustainable well-being". Dasgupta affirms that the exploitation of resources, democracy and protection of the environment should form a "virtuous circle" in synergy: there is more to human development than increased productivity.

Sustainable Development and the Rights of Future Generations

Intergenerational equity as referred to the environment cannot be separated from the issue of so-called "sustainable development" that recurs in the debate on environmental ethics.

The expression leapt to fame in the Brundtland Report (World Commission on Environment and Development 1987) "Sustainable development" is referred to in numerous documents, treaties and declarations (Principle 3 of the "Declaration of Rio", in particular, establishes that "The right to development must be fulfilled so as to equitably meet developmental and environmental needs of present and future generations" (UN 1992)). Sustainable development is also cited as one of the objectives of the UN Millennium Declaration adopted in September 2000 to be achieved by 2015 (UN 2000). States were urged to support environmental sustainability by including the principles of sustainable development in their national development programmes. It is also the focus of the Final Declaration of the UN World Summit on Sustainable Development held in Johannesburg in 2002 (UN 2002).

The aim of the "sustainable development" approach is to overcome both the reduction of development to a merely economic concept and the traditional contrast between the environment and development. The expression gradually acquired the meaning that was already explicit in the Brundtland Report, which was oriented towards demographic control in line with the view that holds man to be more a threat to nature than one of its resources.

Leaving aside these controversial aspects and limiting the debate to intergenerational equity, the notion of "sustainable development" offers at least two important leads. The first is the introduction of an obligatory temporal requisite; states should not form their policies around short-term considerations, but must weigh environmental questions in a long-term perspective. The second concerns responsibility towards all the parties involved, including those as yet unborn.

SOCIAL ASPECTS

Individual Contribution

Some of the difficulties associated with the "rights of future generations" stem from the political and collective dimensions of the problem. There is a deep-seated and widespread perception that individual behaviour can have little significant effect in the long term (despite recognition of the long-term effects of the sum total of a multitude of individual behaviours).

However, this perceived scant usefulness of individual contributions should not translate into an excuse to avoid individual obligations. The tale of the village feast is an effective reminder of the importance of individual actions: each participant was invited to pour a flask of wine into a large barrel. When the feast began and the tap was opened only water came out: each person had thought that just one flask of water in a large barrel of wine would not be noticed.

The attitude of detachedness is also fuelled by the fact that we can directly perceive only the generations in the near future. Those more distantly removed are seen as abstract beings that do not evoke the same participatory involvement we spontaneously feel when thinking of a risk we can perceive directly.

'Globalisation'

If the horizon of the social perspective is expanded it will be seen that there are a number of political problems besides those already mentioned. Without delving deeply into this vast topic, suffice it to recall that in the era of globalisation, restrictive policies intended to avoid despoiling the environment and resources bequeathed to posterity are frequently met with hostility in the less developed nations. These countries often accuse the inhabitants of the industrialised nations of first depleting nature and resources in order to establish their own well-being and then expecting the developing countries to pursue more environment-friendly policies that would deny them the growth potential from which the richer nations have already benefited.

The issue of individual responsibility is also important in the perspective of "globalisation". If individual neglect has contributed much to environmental damage, most of the blame can be traced to the productive processes adopted in all industrialised societies. It follows that when addressing the problems linked to the environment, terms such as global, irreversible, long-term are increasingly encountered. Frequently the scientific facts are unsound or controversial (one example is the different conclusions reached in scientific studies of the greenhouse effect).

The framework of responsibilities is thus very different from that which applies to most aspects of bioethics (and particularly to the clinical field, which is dominated by the doctor/patient twosome with a fairly well defined set of rights and responsibilities). While the values at stake in environmental ethics and law are found on a different level from those proper to medical ethics (the environment is not a person), it must equally be acknowledged that the multiplicity of factors involved means that some of the problems are more complex. It must also be borne in mind that protection of the environment is indispensable to protecting human health and that values that directly affect humans must therefore be considered, not only those that affect the environment as a whole.

The Legal Definition Problem

The legal definition problem has already been mentioned under the heading "common heritage". On the subject of intergenerational equity in general, while there is substantial agreement that each generation has a moral obligation not to penalise succeeding generations by causing irreversible harm to the environment, there are nonetheless contrasting views as to whether this obligation should be defined legally.

There is fairly widespread acceptance of the advisability of including references to future generations in treaties, codes and declarations that are to be respected but which are not legally binding. There are some authors who think that intergenerational equity could effectively be ensured by publishing a "Declaration on the rights of future generations" (Moran-Deviller 2000). Their argument is that only an explicit consecration of these values can ensure that they are effectively heeded in practice. The theoretic procedures required to produce such pronouncements are not simple, as they involve individuals who are not yet born and who thus constitute a completely new legal category. From the cultural viewpoint, too, the issue of intergenerational equity calls for far-sightedness that sees beyond short-term logic and that is itself not without problems.

Although non-binding institutional documents are free of the problems that beset the preparation of texts involving legal obligations, there are still difficulties. One of these is the fact that the different aspects of future generations' rights cannot be properly described without regard to broader problems such as social context, human dignity, living conditions. The importance of these aspects leads many to insist that intergenerational equity should not be limited to tacit or implicit definitions: the rights of future generations should, in this view, be explicitly included at various legal levels, in both national (constitutions and regulatory provisions) and supranational (multilateral agreements, binding conventions, "soft laws") documents.

Yet others sustain that if future generations cannot claim their own rights, special institutions should be set up with the task of defending their interests. This approach would provide some sort of albeit incomplete and unusual vicarious "participation" by future generations in lieu of their direct involvement.

SOME ANSWERS

The problems of intergenerational justice in the matter of environmental protection inevitably pose questions concerning the value of solidarity with persons who are temporally remote. This issue has been addressed by sociologists, philosophers and ethicisits, and has been amply dealt with in the literature (see, for example, Dieter Birinbacher: "La responsabilité envers les générations futures" (Birinbacher 1994)).

The views of some of the authors who have written on this topic are briefly summarised below.

Bryan G. Norton

According to Bryan G. Norton, the two aforementioned problems, relating to identity and temporal remoteness, can be overcome if we abandon the individualistic approach to ethics (Norton 1987) and adopt Edmund Burke's criticisms of the social contracts proposed by Hobbes, Locke and Rousseau. Burke writes of a partnership "not only between those who are living, but between those who are living, those who are dead, and those who are to be born. Each contract of each particular state is but a clause in the great primeval contract of eternal society, linking the lower with the higher natures, connecting the visible and the invisible world, according to a fixed compact sanctioned by the inviolable oath which holds all physical and all moral natures, each in their appointed place contract" (Burke 1790, quoted in Carpenter 1998, p. 275-293). Norton thus proposes a theory that brings together elements that are distant in both time and space, although none is explicitly identified.

Edith Brown Weiss

Edith Brown Weiss proposes another approach that underlines the importance of overcoming individualism and recognising the rights of future generations. Brown Weiss examines the meaning of mankind's "common patrimony" referred to not only in the major religions but also in numerous cultural and political movements. There is an obligation to defend the planetary systems that sustain life, ecological processes, environmental conditions and the cultural resources necessary for the survival and well-being of the human species (Brown Weiss 1988, p. 23).

Weiss uses the expression "planetary stewardship", for the preservation and transmission of the "common patrimony" (Brown Weiss 1988, pp. 2-3). She proposes four principles for the realisation of "planetary stewardship":

- to encourage intergenerational equity by banning inconsiderate exploitation, but at the same time to avoid this being translated into unreasonable obligations to protect future generations;
- to avoid all attempts to predict or quantify the value of future generations;
- to choose principles that are reasonable, clear and practicable;
- to choose values that are generally shared by different cultural traditions and that are acceptable to different economic and social systems.

Weiss further shows how *inter*generational equity can only be achieved if we also cultivate *intra*generational equity. To this end, she proposes:

- conserving options: the conditions bequeathed by each generation to the next should be such that no option is definitively precluded;
- conserving quality: the condition in which each generation leaves the planet should not be worse than that in which they found it;

- conserving access: actions intended to benefit future generations but which may involve excessive burdens for particular groups of present generations (e.g. developing countries) should be avoided.

Positions similar to those expounded by Brown Weiss have met with some support, given that the general principles they propose are acceptable in widely divergent cultural backgrounds, and because they treat the issue of intergenerational equity as separate from concern for single individuals.

Aldo Leopold

The "land ethic" proposed by Aldo Leopold also refers to intergenerational equity: "A thing is right when it tends to preserve the integrity, stability, and beauty of the biotic community. It is wrong when it tends otherwise" (Leopold 1949). In a strongly anti-anthropocentric stance, Leopold insisted on the concept of community, in the sense of forming a whole with all the components of nature, and considered that the only legitimiate behaviour is that which does not interfere with the equilibrium between different components.

Henri Skolimowski

Henri Skolimowski used the expression "reverential development based on ecological values", echoing the better known "reverence for life" used by Albert Schweitzer (Schweitzer, and Roy. 1947). Skolimowski writes that "in proposing a new form of development, reverential development based on ecological values, I intend to promote simultaneously the sustainability of the planet, the dignity of different people and the unity of humanity, now divided by inappropriate development". The "reverential development" proposed by Skolimowski aims to: (a) combine the economic with the ethical and the reverential; (b) combine contemporary ethical imperatives with traditional ethical codes; (c) attempt to serve all people of all cultural traditions; (d) promote peace between men and nature" (Skolimowski 1990, pp. 97-103).

John Rawls

The proposals of the writers mentioned above were specifically concerned with the problem of the environment.

Given the breadth of the debate stirred up from the 1960s onwards in various fields (political, social, philosophical), it is as well to recall the contribution of John Rawls (Korsgaard, and Freeman 2001, pp. 1454-61). Some of his general arguments have been studied with specific reference to the field of intergenerational environmental equity.

Rawls initiates his analysis from an "original position", a situation perceived from behind "a veil of ignorance". From this position certain fundamental values or principles that are innate in rational beings can be established. In other words, the principles should be decided by individuals who meet prior to laying the foundations of a civilised society and who are unaware of their own social status within it. In this way, according to Rawls, freedom, equality and justice would be identified as the principles that govern a fair society (Rawls 2001).

Rawls defines justice "the first virtue of social institutions, as truth is of systems of thought". He interprets equity in particular as a collective well-being, and draws on both liberal democracy and systems with a strong social state. He proposes uniting in a single system the equality propounded by socialist systems and the freedom propounded by liberal systems.

Rawls' moral rule is based on "agreement" and "fairness". In the procedural ethics derived from these, that which is right is the result of a consensual decision (Rawls 2002): in other words, ethics are case-sensitive and constantly question the values to which they refer.

Obviously, future generations cannot be parties to this contract directly. Nonetheless, the approach of procedural ethics had a profound influence on the debate about intergenerational equity.

Conclusion

The Importance and Topicality of the Problem

It is evident that when faced with environmental problems in a temporal perspective, projected towards the future, the idea of "patriotism" limited to considering only one's own back yard becomes increasingly unrealistic.

It is equally evident that terms such as global, irreversible, long-term effects and risk uncertainty are called for when addressing environmental problems. To deal with these concepts, bioethics must draw on a multitude of disciplines, given that neither "bedside" or "laboratory" bioethics are sufficient.

In October 2004 the Pontifical Council for Justice and Peace published the "Compendium of the Social Doctrine of the Church". Paragraph 467 states: "Responsibility for the environment, the common heritage of mankind, extends not only to present needs but also to those of the future. We have inherited from past generations, and we have benefited from the work of our contemporaries: for this reason we have obligations towards all, and we cannot refuse to interest ourselves in those who will come after us, to enlarge the human family. The reality of human solidarity brings us not only benefits but also obligations". "This is a responsibility that present generations have towards those of the future, a responsibility that also concerns individual States and the international community" (Pontifical Council for Justice and Peace 2004, par. 467).

Operational Criteria

We cannot predict the future. One could therefore object that there is a contradiction in burdening contemporary generations with responsibility towards a future we are ill-equipped to foresee. The most we can do, perhaps, is to say that if current behaviour continues into the future certain consequences will presumably occur. It is more difficult to predict the consequences of actions that are not being taken.

These uncertainties lend controversy to the notion of "responsibility towards future generations" but cannot justify a refusal to acknowledge that responsibility.

In seeking to identify the salient points to consider in situations of uncertainty, two elements must be emphasised, in particular: the duty not to cause irreversible harm to the environment and the duty not to narrow the scope of options (such as biological diversity, or other forms) available to future generations.

Focusing on future generations can help to broaden our horizons, to concentrate less on matters of immediate concern and to be aware that there may be other ways of acting. In this sense it can be a stimulus to awareness and an expansion, rather than a limitation of freedom.

It should also be noted that, because the only way present generations can act with respect for the rights of future generations is to do something in the present, there are some who hold that concern for the future could, in some cases, be used as a pretext to choose one set of options over another in pursuit of an immediate objective.

On the Relationship between Intergenerational Equity and the Rights of Contemporaries

The key to protecting the rights of future generations is not to deplete environmental resources irreversibly. In situations of particularly serious need this could conflict with the fundamental rights of contemporary populations. Just as it must be recognised that today's populations are not worth more than tomorrow's, so must it be recognised that neither are they worth less. Criteria must therefore be identified to permit a fair compromise. This unquestionably poses particular problems for groups of people in economically backward nations whose survival is closely tied to the exploitation of local resources (e.g. forests).

On Freedom and on Duty

The term "freedom" is constantly evoked in the debate on intergenerational equity. It is emphasised that "freedom" to choose must be passed on intact to future generations, and that the common heritage must be a good that can conserve in the future the potential to adapt to uses not foreseeable in the present (De Montgolfier, Natali, and Méhaignerie 1987). It is perhaps worth pointing out that the transmission of a heritage involves the transmission not only of freedom, but also of a duty. Future generations are thus burdened with the same responsibility.

References

Assemblée Nationale de la République Française. 1983. Loi n°83-8 du 7 janvier 1983 Relative à la Répartition de Compétences entre les Communes, les Départements, les Régions et l'Etat (Loi Defferre). *Journal Officiel de la République Française* 9 Janvier 1983: 215-30.

Birinbacher, Dieter. 1994. *La Responsabilité Envers les Générations Futures*. Paris: Presses Universitaires de France.

Birnie, Patricia; Boyle, Alan; Redgwell Catherine. 2002. *International Law and the Environment*. Oxford: Oxford University Press.

Brown Weiss, Edith. 1988. *In Fairness to Future Generations*.. Tokyo and New York: University and Transnational Publishers.

Burke, Edmund. 1790. Reflections on the French Revolution. Quoted in: Carpenter, Stanley Robert. 1998. Sustainability. In *Encyclopedia of Applied Ethics*, ed. Ruth F. Chadwich, Vol. 4, pp. 275-93. San Diego: Academic Press.

Butera, Federico M. 2001. La Comunità Scientifica Internazionale e l'Ambiente. In *Etica Ambiente Sviluppo. La Comunità Internazionale per una Nuova Etica dell'Ambiente*, ed. Amedeo Postiglione and Antonio Pavan, pp. 75-134. Napoli: Edizioni Scientifiche Italiane.

Buttel, Fredrick H. 1986. Sociologie et Environnement: la Lente Maturation de l'Écologie *Humaine. Revue Internationale des Sciences Sociales* 109: 359-77.

Dasgupta, Partha. 2001. *Human well-being and the natural environment*. Oxford: Oxford University Press.

De Marchi, Bruna; Pelizzoni, Luigi; and Ungaro, Daniele. 2001. *Il Rischio Ambientale*, pp. 11-36. Bologna: Il Mulino.

De Montgolfier, Jean; Natali, Jean-Marc; and Méhaignerie, Pierre. 1987. *Le Patrimoine du Future : Approches pour une Gestion Patrimoniale des Ressources Naturelles*. Paris: Economica.

Dower, Nigel. 1998. Development Issues, Environmental. In: *Encyclopedia of Applied Ethics*, ed. Ruth F. Chadwick, Vol. 1. pp. 767-79. San Diego: Academic Press.

Durkheim, Émile. 1895. *Les Règles de la Méthode Sociologique*. Paris: Alcan

European Commission - Foresight to the New Technology Wave Expert Group. 2004. *Converging Technologies. Shaping the Future of European Societies*. Bruxelles: European Commission. Available at: ec.europa.eu/research/conferences/2004/ntw/pdf/final_report_en.pdf [Accessed 12 July 2009].

Goffi, Jean-Yves. 2001. Jonas Hans. In: *Dictionnaire d'Éthique et de Philosophie Morale*, ed. Monique Canto-Sperber, pp. 818-9. Paris: Presses Universitaires de France.

Grima, George. 2004. Giustizia Intergenerazionale. In: Nuovo *Dizionario di Bioetica*, ed Salvatore Privitera and Salvino Leone, pp. 546-50. Roma and Acireale: Città Nuova and Istituto Siciliano di Bioetica.

Humbert, Geneviève. 1992. À Chacun son Patrimoine ou Patrimoine Commun? In: *Sciences de la Nature, Sciences de la Société. Les Passeurs de Frontière*, ed. Marcel Jollivet, pp. 187-96. Paris: Éditions du CNRS.

Jonas, Hans. 1979. *Das Prinzip Verantwortung. Versuch einer Ethik für die technologische Zivilisation*. Frankfurt a. M.: Insel Verlag.

Kant, Immanuel. 1993. *Grounding for the Methaphysics of Morals*. Indianapolis: Hackett (Original title: Kant, Immanuel. 1785. Grundlegung zur Metaphysik der Sitten).

Korsgaard, Christine M; and Freeman, Samuel. 2001. Rawls, John. In *Encyclopedia of ethics*, ed. Lawrence C. Becker and Charlotte B. Becker, pp. 1454-61. New York: Routledge - Taylor & Francis.

Leopold, Aldo. 1949. *A Sand County Almanac and Sketches Here and There*. New York: Oxford University Press.

Morand-Deviller, Jacqueline. 2000. *Le Droit de l'Environnement*. Paris: Presses Universitaires de France.

Norton, Bryan G. 1982. Environmental Ethics and the Right of Future Generations. *Environmental Ethics* 4 (winter): 319-35.

Norton, Bryan G. 1987. *Why Preserve Natural Variety?* Princeton: Princeton University Press.

Ost, François. 1995. *La Nature et la Loi. L'Écologie à l'Épreuve du Droit*. Paris: La Découverte.

Passmore, John. 1974. *Man's Responsibility for Nature*. New York: Scribners.

Pontifical Council for Justice and Peace. 2004. *Compendium of the Social Doctrine of the Church*, par. 467. Vatican City: Vatican Publishing House.

Potter, Van Rensselaer. 1970. Bioethics: the Science of Survival. *Perspectives in Biology and Medicine* 1: 127-53.

Potter, Van Rensselaer. 1971. *Bioethics, Bridge to the Future*. Englewood Cliffs, NJ: Prentice-Hall Inc.

Rawls, John. 1971. *A theory of justice*. Cambridge, MA: Belknap Press.

Rawls, John. 2001. *Justice as Fairness: a Restatement*. Cambridge, MA: Belknap Press.

Roussopoulos, Dimitrios I. 1994. *L'écologie politique*. Montréal - Cap-Saint-Ignace: Écosociété.

Schweitzer, Albert; and Roy, Charles. 1947. *Albert Schweitzer: An Anthology*. New York and London: Harper & Brothers.

Skolimowski, Henri. 1990. Reverence for life. In: *Ethics of Environment and Development: Global Challenges, International Response*, ed. J. Ronald Engel and Joan Gibb Engel, p. 97-103. Tucson: University of Arizona Press.

Thomas, Yves. 1980. Res, chose et patrimoine. Note sur le rapport sujet-objet dans en droit romain. *Archives de Philosophie du Droit*. Quoted in: Ost, François. 1995. *La Nature et la Loi. L'Écologie à l'Épreuve du Droit*. Paris: La Découverte.

United Nations. 1982. *United Nations Convention on the Law of the Sea (UNCLOS)*. Montego Bay, 10 December 1982. Available at: www.un.org/depts/los/convention_agreements/texts/unclos/closindx.htm [Accessed 12 December 2009].

United Nations. 1992. *Conference on Environment and Development. Rio Declaration on Environment and Development*. U.N. Doc. / CONF.151/26/, 1992. Available at: www.unep.org/documents.multilingual/default.asp?documentid=78&articleid=1163 [Accessed 12 December 2009].

United Nations. 2000. *United Nations Millennium Declaration. Resolution Sdopted by the General Assembly. Fifty-fifth Session. Agenda item 60(b)*. A/RES/55/2, 2000. Available at: www.un.org/millennium/declaration/ares552e.htm [Accesses 12 December 2009].

United Nations. 2002. *The Johannesburg Declaration on Sustainable Development*. Available at: *www.un.org/esa/sustdev/documents/wssd_poi_pd/english/poi_pd.htm* [Accessed 12 December 2009].

World Commission on Environment and Development (Brundtland, Gro Harlem). 1987. *Our Common Future*. Oxford: Oxford University Press.

In: Bioethics: Issues and Dilemmas
Editor: Tyler N. Pace
ISBN: 978-1-61728-290-4

Chapter 5

TWO DIFFERENT APPROACHES TO PRINCIPLES OF BIOMEDICAL ETHICS: A PHILOSOPHICAL ANALYSIS AND DISCUSSION OF THE THEORIES OF THE AMERICAN ETHICISTS TOM L. BEAUCHAMP AND JAMES F. CHILDRESS AND THE DANISH PHILOSOPHERS JAKOB RENDTORFF AND PETER KEMP

Mette Ebbesen
Centre for Bioethics and Nanoethics, University of Aarhus, Denmark

1. INTRODUCTION

Most research within biomedical ethics consists of theoretical reflections regarding which ethical theories or principles are useful to analyse ethical issues in the field of biomedicine. The theories of the American ethicists Tom L. Beauchamp & James F. Childress (2009) and the Danish philosophers

Jakob Rendtorff & Peter Kemp (2000) are examples[1]. Beauchamp & Childress examined considered moral judgements and the way moral beliefs cohere and found that the general principles of beneficence, nonmaleficence, respect for autonomy, and justice play a vital role in biomedical ethics (Beauchamp & Childress, 2009, p. 13). These authors believe that the four principles are not only specific for biomedical ethics, they are found in all cultures in everyday life because they are part of a cross-cultural common morality shared by all persons committed to morality (Beauchamp & Childress, 2009, p. 4). The four clusters of principles provide a framework of norms to start with in biomedical ethics. As a starting point no principle is weighted higher than the other principles. When occasion arises, the principles are weighted, balanced, and specified. As will become clear, Rendtorff & Kemp do not find the approach of Beauchamp & Childress convincing, and they developed an alternative theory based on the following four basic ethical principles: autonomy, dignity, integrity, and vulnerability. Rendtorff & Kemp regard Beauchamp & Childress' approach as individualistic, and they believe that their alternative fills out an empty space in the American theory by protecting "the fragile and finite, bodily incarnated human person" and thereby leads to a wider view of the human person (Rendtorff & Kemp, 2000, p. 314).

First, this article presents a philosophical analysis of the bioethical theory of Beauchamp & Childress. Next, the theory of Beauchamp & Childress is discussed and compared it with the theory of Rendtorff & Kemp.

2. The Four Principles of Biomedical Ethics of Tom L. Beauchamp and James F. Childress

The Hippocratic writers were among the first systematic medical ethics writers in the Western tradition. Based on the Hippocratic oath, which expresses an obligation of beneficence and an obligation of nonmaleficence, the Hippocratic tradition focused on promoting the good and reducing harm: "I will use treatment to help the sick according to my ability and judgment, but I will never use it to injure or wrong them" (Beauchamp & Childress, 2009, p. 149). As Beauchamp & Childress emphasise, the Hippocratic tradition ignored ethical issues of "truthfulness, privacy, the distribution of health care

1 It should be noted, however, that the four clusters of principles presented by Beauchamp & Childress do not constitute a general ethical theory but only a framework of norms to start with in biomedical ethics (Beauchamp & Childress, 2009, p. 16).

resources, communal responsibility, the use of research subjects, and the like" (Beauchamp & Childress, 2009, p. 1). Currently, bioethical researchers are trying to develop broader ethical theories to cover these ethical issues. For instance, Beauchamp & Childress think that general reflections on respecting the autonomy of the patient or human subject and considerations about justice with regard to health care allocation must supplement reflections on the obligations to promote good and reduce harm. By examining considered moral judgments within biomedicine, Beauchamp & Childress found that these judgements are mainly justified by four basic ethical principles which are central to and play a vital role in biomedical ethics. These principles are respect for autonomy, nonmaleficence, beneficence, and justice (Beauchamp & Childress, 2009, p. 13). In figure 1, a brief formulation of the bioethical principles of Beauchamp & Childress is presented.

The Principle of Respect for Autonomy
• "As a negative obligation: Autonomous actions should not be subjected to controlling constraints by others" (Beauchamp & Childress, 2009, p. 104). • "As a positive obligation, this principle requires both respectful treatment in disclosing information and actions that foster autonomous decision making" (Beauchamp & Childress, 2009, p. 104). Furthermore, this principle obligates to "disclose information, to probe for and ensure understanding and voluntariness, and to foster adequate decision making" (Beauchamp & Childress, 2009, p. 104).
The Principle of Beneficence
• One ought to prevent and remove evil or harm • One ought to do and promote good (Beauchamp & Childress, 2009, p. 151).
The Principle of Nonmaleficence
"One ought not to inflict evil or harm", where harm is understood as "thwarting, defeating, or setting back some party's interests" (Beauchamp & Childress, 2009, pp. 151-152).
The Principle of Justice
Beauchamp & Childress do not present one principle of justice but a concept that can be determined in various ways. They adopt the egalitarian theory by John Rawls where justice means fair opportunity: The goods to be distributed are compensations for disadvantages caused by the natural or social "lottery". Thus fair opportunity means that a person born disabled should receive special services. Beauchamp & Childress think that a fair health care system includes two strategies for health care allocation: 1) a utilitarian approach emphasising maximal benefit to patients and society, and 2) an egalitarian strategy emphasising the equal worth of persons and fair opportunity. They defend the egalitarian view point that all citizens enjoy a right to a decent minimum of health resources (Beauchamp & Childress, 2009, pp. 240-281).

Figure 1. The four principles of biomedical ethics. A brief formulation of the four ethical principles of Beauchamp & Childress (2009): respect for autonomy, beneficence, nonmaleficence, and justice.

Prima Facie Binding, Balancing, and Specification

Beauchamp & Childress were inspired by the Scottish philosopher W. D. Ross, who, in *The Right and the Good* (1930), claims that common sense tells us that we have some prima facie duties (also called conditional duties) to do special acts (e.g. keeping a promise) (Ross, 1930, pp. 40-41). Like Ross, Beauchamp & Childress consider the bioethical principles prima facie binding, that is, they are binding in any situation if they do not conflict with other prima facie principles (Beauchamp & Childress, 2009, p. 15). They claim that each prima facie principle has weight without assigning a priority weighting or ranking. Which principle to set aside in a case of conflict within biomedical practice depends on the concrete situation. However, Beauchamp & Childress write that it is not easy to balance the principles in an actual situation (Beauchamp & Childress, 2009, p. 16). Many critics claim that in the first editions of *Principles of Biomedical Ethics,* they lack a rule for how to weigh up principles if they come into conflict (DeGrazia, 1992; Holm, 1995; Strong, 2000). In the third edition of their book, Beauchamp & Childress are aware of the problem of balancing. They specify conditions that should be fulfilled in order to allow one prima facie principle to weigh higher than another:

"The following are all requirements for justified infringements of a prima facie principle or rule:

[1] the moral objective justifying the infringement must have a realistic prospect of achievement;
[2] infringement of a prima facie principle must be necessary in the circumstances, in the sense that there are no morally preferable alternative actions that could be substituted;
[3] the form of infringement selected must constitute the least infringement possible, commensurate with achieving the primary goal of the action; and
[4] the agent must seek to minimize the effects of the infringement" (Beauchamp & Childress, 1989, p. 53).

In the fifth edition of their book, Beauchamp & Childress sharpen these conditions that must be fulfilled for one prima facie principle to outweigh another. They add the following premise: "The agent must act impartially in regard to all affected parties; that is, the agent's decision must not be influenced by morally irrelevant information about any party" (Beauchamp &

Childress, 2001, p. 20). Thereby, Beauchamp & Childress introduce the demand of impartiality.

However, Henry S. Richardson claims that the balancing metaphor should be exchanged for a specification of principles (Richardson, 2000). A specification of a principle is an interpretation of an original abstract principle, which is consistent with the original principle. “Specification is the progressive filling in of the abstract content of principles, shedding their indeterminateness, and thereby providing action-guiding content” (DeGrazia & Beauchamp, 2001, p. 34). In the fifth edition of their book, Beauchamp & Childress emphasise that bioethical principles need to be specified so that they approach the specific problem, and they make Richardson’s theory of specification part of their method (Beauchamp & Childress, 2001, pp. 15-18).

The sensitiveness of Beauchamp & Childress’ theory can be illustrated by the changes made since the first editions of their theory in 1979 to the sixth edition in 2009. They have taken their critics into account during the last 30 years by incorporating their comments and suggestions and simultaneously publishing papers in which they discuss their theory (Beauchamp, 1995; Beauchamp, 2000; Beauchamp, 2003). Some of the critique of the first edition of the method is that they emphasise theory more than practice and that they stress procedure more than practical wisdom (Strong, 2000; King & Churchill, 2000). Beauchamp & Childress answer these points of critique in the fifth and sixth edition of their work by combining their principlism with the ethics of virtue and the coherence theory described below in the next section.

Justification of the Principles

In the first editions of *Principles of Biomedical Ethics*, Beauchamp & Childress defend a top-down model of justification by arguing that ethical judgements and principles are justified deductively from ethical theories (Beauchamp & Childress, 1989, pp. 3-23). In later editions of their book, they emphasised that the basic principles can be drawn from a common cross-cultural morality (Beauchamp & Childress, 2001, pp. 12, 23). Beauchamp & Childress write: "“The top” (principles, theories) and “the bottom” (cases, individual judgements) are not solely sufficient for biomedical ethics. Neither general principles nor paradigm cases have sufficient power to generate conclusions with the needed reliability” (Beauchamp & Childress, 2001, p. 397). Based on this, they defend the so-called “Integrated Model: The

Coherence[2] Theory" inspired by the American political philosopher John Rawls.

The coherence theory, also called reflective equilibrium, was given prominence by Rawls in his book *A Theory of Justice* (1971) as a method for developing and justifying principles for a just society. Rawls focuses on a system with three levels: particular moral judgements, first principles, and general convictions. He states that particular moral judgements are justified by the overall coherence of this system, and he uses the concept of wide reflective equilibrium to describe this state of coherence (Rawls, 2001). Rawls writes that "justification is a matter of the mutual support of many considerations, of everything fitting together into one coherent view" (Rawls, 1971, p. 19). To reach wide reflective equilibrium, we begin by screening our initial moral judgements to get rid of those in which we have little confidence. We then search for general moral principles, justifying the remaining as considered moral judgements. We may find reason to revise or discard some of the considered moral judgments that conflict with trustworthy moral principles. Rawls believes that the process of comparing principles with considered judgments leads us to go back and forth: revising our principles and our considered moral judgements until the principles match, fit, or are in line with our considered moral judgements, and consistency is reached. Finally, we have to subject the moral principles we arrive at to alternative moral conceptions and perspectives and to the force of various arguments for them. Wide reflective equilibrium is achieved when our considered judgements match or are in line with our general principles duly pruned and adjusted. However, we never assume a completely stable, wide reflective equilibrium. For instance, this equilibrium is liable to be upset by particular cases, which may lead us to readjust our judgments or principles (Rawls, 1971, pp. 18, 42-45; Rawls, 2001, pp. 29-33). Hence, the adjusting goes on continually.

2 The notion of "coherence" is not well defined by Beauchamp & Childress. Philosophers agree that coherence is not only characterised by consistency (Petersson, 1998, pp. 127-134; Kappel, 2006). According to the Danish philosopher, Klemens Kappel, coherence is characterised by consistency, systematicity (a belief set should contain explanatory relations), generality (a belief set should contain general beliefs that cover a larger area rather than a smaller one), and simplicity (general explanatory beliefs should be few and simple rather than many and complex) (Kappel, 2006).

Coherence - Not Enough for Justification

In the latest edition of their book, Beauchamp & Childress use wide reflective equilibrium as basic methodology (Beauchamp & Childress, 2009, p. 385). As shown above, according to the theory of reflective equilibrium, justification is a reflective testing of our moral judgements, moral principles, theoretical postulates, and other relevant moral beliefs to make them as coherent as possible (Beauchamp & Childress, 2009, p. 382). However, Beauchamp & Childress emphasise that “although justification is a matter of reflective equilibrium in this model, bare coherence never provides a sufficient basis for justification, because the body of substantive judgments and principles that cohere could themselves be morally unsatisfactory” (Beauchamp & Childress, 2009, p. 384). They show this by referring to the “Pirates’ Creed of Ethics” or “Custom of the Brothers of the Coast”, a coherent set of rules formed by pirates in 1640 governing mutual assistance in emergencies, penalties for prohibited acts, etc. This system of rules and principles is coherent, but it involves immoral activities such as the awarding of slaves as compensation for injury (Beauchamp & Childress, 2009, pp. 384-385). By using the example of the Pirates’ Creed of Ethics, Beauchamp & Childress illustrate that coherence is not a sufficient condition for moral judgements to be justified by ethical principles since the system, although coherent, does not constitute an acceptable code of ethics. Beauchamp & Childress believe that considered judgements in the coherent system should have a rich history of moral experience that undergirds confidence that they are credible and trustworthy (Beauchamp & Childress, 2009, p. 385). Therefore, the method of reflective equilibrium needs to be combined with the common morality theory (Beauchamp & Childress, 2009, pp. 384-385).

The Common Morality

The four principles of respect for autonomy, beneficence, nonmaleficence, and justice are part of a common cross-cultural morality: “The four clusters of principles derive from considered judgements in the common morality and professional traditions in health care, particularly medicine and nursing” (Beauchamp & Childress, 2009, p. 25). By the term “common morality” Beauchamp refers to “the set of norms shared by all persons committed to the objectives of morality. The objectives of morality, I will argue, are those of promoting human flourishing by counteracting conditions that cause the quality of people's lives to worsen" .. "It is applicable to all persons in all places, and all human conduct is rightly judged by its standards. Virtually all people in all cultures grow up with an understanding of the basic demands that morality makes upon everyone. They know not to lie, steal, break promises, and the like" (Beauchamp, 2003). Beauchamp & Childress write that since many amoral persons do not care about the moral obligations and demands, it would be ridiculous to suppose that these common morality norms are accepted by all persons. Nonetheless, they believe that all persons in all cultures who take moral conduct seriously do accept the norms of the common morality (Beauchamp & Childress, 2009, p. 4).

The common morality plays a foundational role in the latest editions of Beauchamp & Childress' theory (Beauchamp & Childress, 2009, p. 388). They write that no more basic moral content exists than the body of rules and judgements developed from the four clusters of principles. Furthermore, they state that the common morality can function as a basis for evaluation of groups whose moral conduct seems in some respect deficient (Beauchamp & Childress, 2009, p. 387).

Empirical Perspective on the Common Morality

Beauchamp & Childress not only justify the four principles normatively by the common morality, they also justify them empirically by arguing that the common morality *describes* what all people actually believe (Beauchamp & Childress, 2001, p. 4). They do not claim that the moral norms of all societies (specific moralities) are indistinguishable; rather they state that the most general norms (e.g. the four principles) are held in common (Beauchamp & Childress, 2009, p. 392). Beauchamp & Childress claim that it can be tested empirically whether the four bioethical principles are important to biomedical ethics and whether they are part of a common morality (Beauchamp, 2003; Beauchamp & Childress, 2009, pp. 392-394). However, they do not present any empirical data generated systematically by qualitative research methods to support this position. Beauchamp & Childress are inspired by Ross with regard to their choice of principles. Ross regards the duties of fidelity, reparation, gratitude, justice, beneficence, self-improvement, and nonmaleficence as prima facie duties (Ross, 1930, pp. 21-22). These prima facie duties are objective facts involved in the nature of situations in human practice, therefore they are self-evident and do not need to be proven (Ross, 1930, pp. 19-21, 29-30). Beauchamp & Childress' choice of principles is almost in agreement with

Ross' prima facie duties. However, ethical considerations unrecognized by Beauchamp & Childress' theory may be brought to light by empirical studies of the ethical reasoning of health care professionals and researchers (Ebbesen & Pedersen, 2007[a], 2007[b], 2008[a], 2008[b]).

3. The Basic Ethical Principles of Jacob Rendtorff and Peter Kemp

Rendtorff & Kemp regard their ethical theory as an alternative to Beauchamp & Childress' principles, with a considerably different approach (Rendtorff & Kemp, 2000, pp. 18-19, 314). They formulated four basic ethical principles of autonomy, dignity, integrity, and vulnerability that present a normative framework for the protection of the human person in biomedical development (Rendtorff, 2002). According to Rendtorff (2002), the principles are institutionalised in various ways in the different European countries. One example is in biomedical legislation. Rendtorff & Kemp (2000) performed an empirical analysis in which they explored the principles in legal documents in different European countries. Based on this analysis, they define the principles as "middle level" principles "between visions of the good life for the human person and norms for concrete cases of application" (Rendtorff, 2002).

Rendtorff & Kemp performed what they call a hermeneutic analysis of the possible uses of the ethical principles, and they "justify the principles within a phenomenology of moral values in human intersubjective relations" (Rendtorff, 2002). The principles are legal and ethical guidelines for the physician-patient relationship in the clinic. Rendtorff (2002) believes that respect for these principles will contribute to a better physician-patient communication.

As described above, Rendtorff & Kemp regard their approach as a European alternative to Beauchamp & Childress' theory. Rendtorff & Kemp state that Beauchamp & Childress have a minimalist conception of the human person that regards autonomy as the only guiding principle (Rendtorff & Kemp, 2000, pp. 18-31). They write that in Beauchamp & Childress' theory "autonomy remains merely an ideal because of its structural limitations, i.e. human dependence on outer factors, lack of information, reduced capacity of reasoning, etc." (Rendtorff, 2002). They believe that "this limitation of the concept of autonomy is also due to tension between the human existence as an unencumbered self and the embodied, embedded, character of human experience" (Rendtorff, 2002). They do not think that autonomy is a sufficient normative concept to protect humans in ethical and legal affairs, because some persons may not have the capacity to reason or this capacity may be limited. According to Rendtorff & Kemp, this could be the case for children, senile people, insane people, etc, and also for normal, intelligent people who feel weak and dependent on others, or do not understand the scientific project they are asked to participate in (Rendtorff & Kemp, 2000, pp. 18-31; Rendtorff, 2002). Therefore, Rendtorff & Kemp believe that other principles such as dignity, integrity, and vulnerability must supplement the principle of autonomy to protect the human person in biomedicine (Rendtorff & Kemp, 2000, pp. 18-31). In figure 2, the basic ethical principles of Rendtorff & Kemp are presented.

Autonomy
Five important meanings of autonomy can be put forward:
• "The capacity for the creation of ideas and goals for life" • "The capacity of moral insight" • The capacity of ""Self-legislation" and privacy" • "The capacity of rational decision and action without coercion" • "The capacity of informed consent to medical experiments" (Rendtorff & Kemp, 2000, p. 25).
Dignity
The concept of dignity has two important dimensions. Originally, "it expressed an intersubjective recognition of a distinct characteristic or aspect of personality. In that sense it is the quality of being a worthy or honourable person in society". Secondly, dignity becomes a characteristic "that every human being has as such, requiring that we must respect our fellow human being as a bearer of rights and duties. In this context dignity signifies a substantial aspect and the intrinsic value of the humanity of the person" (Rendtorff & Kemp, 2000, pp. 31-32).
Integrity
The definition of integrity includes the following moral dimensions:
• "Integrity as a created and narrated coherence of life, as a wholeness and completeness of a life story, that must not be violated" • "Integrity as a personal sphere for experiences, creativity and personal self-determination (autonomy)" (Rendtorff & Kemp, 2000, p. 40).
Vulnerability
"The temporal and finite quality of all human life indicates that the human condition is very fragile". "The influence of mortality and destiny on human life cannot be ignored. Vulnerability means that we have to live with morality and take care of the other as a fragile situated subject". "Vulnerability is important as the foundation of the notions of care, responsibility and empathy with the other" (Rendtorff & Kemp, 2000, p. 49).

Figure 2. The basic ethical principles of Rendtorff & Kemp. A brief formulation of the basic ethical principles of Rendtorff & Kemp (2000): autonomy, dignity, integrity, and vulnerability.

4. Comparison and Discussion of the Ethical Principles of Beauchamp and Childress and Rendtorff and Kemp

The theories of Beauchamp & Childress and Rendtorff & Kemp are developed from different philosophical traditions. Beauchamp & Childress base their theory in the analytical Anglo-Saxon tradition, which stresses the human being as rational. Rendtorff & Kemp, on the other hand, are inspired by French philosophy and develop their theory within a phenomenological and hermeneutical frame of reference, which values the human being as a fragile situated subject in intersubjective relations. Rendtorff & Kemp describe their ethical theory as including four principles of autonomy, dignity, integrity, and vulnerability. However, ethical principles contain an obligation, which for instance can be formulated as "you ought to respect ...". What Rendtorff & Kemp call principles do not contain obligations. Hence, strictly speaking, they cannot be considered as principles but as ethical concepts which can be reformulated into ethical principles in the following way: respect for autonomy, respect for dignity, respect for integrity, and respect for vulnerability. By this reformulation, the principle of respect for autonomy, for instance, contains the following obligation: "you ought to respect the autonomy of persons". So to be able to compare the theories of Beauchamp & Childress and Rendtorff & Kemp, the ethical concepts of Rendtorff & Kemp need to be reformulated into principles[3].

In the following, the principles of Rendtorff & Kemp and Beauchamp & Childress are compared. Beauchamp & Childress divide their principle of respect for autonomy into a negative and a positive obligation. The negative obligation says that autonomous actions should not be subjected to controlling constraints by others, and the positive obligation requires respectful treatment in disclosing information and actions that foster autonomous decision making. Furthermore, the positive obligation embodies a command to disclose information, to probe for and ensure understanding and voluntariness, and to foster adequate decision making (Beauchamp & Childress, 2009, p. 104). The positive obligation of Beauchamp & Childress' theory is not seen in the theory of Rendtorff & Kemp. However, the negative obligation is in line with three of the five meanings of autonomy of Rendtorff & Kemp: the capacity for self-

[3] In the rest of this article, however, Rendtorff & Kemp's notions are used to describe their principles even though stringently they cannot be described as principles in their current form.

legislation and privacy, the capacity for rational decision and action without coercion, and the capacity for giving informed consent to medical experiments (Rendtorff & Kemp, 2000, p. 25). Hence, Beauchamp & Childress' principle of respect for autonomy is broader than Rendtorff & Kemp's principle of autonomy because it contains both a negative and a positive obligation.

Because some persons may not have the capacity to reason or this capacity may be limited, Rendtorff & Kemp do not think that autonomy is sufficient to protect human beings with regard to biomedical development. Therefore, the principle of autonomy needs to be supplemented with the principles of dignity, integrity, and vulnerability (Rendtorff & Kemp, 2000, pp. 18-31).

Beauchamp & Childress' theory does not include a principle on human dignity. According to Beauchamp, he and Childress place "much of the content found in various - and often competing - conceptions of human dignity in other principles and rules. Most obviously, the principle of respect for autonomy expresses part of what is associated with human dignity. The principle of nonmaleficence includes not thwarting basic goods such as life and not inflicting what is often called dignitary harm. In addition, the rule of privacy captures another aspect of human dignity, while both respect for autonomy and justice rule out objectification and instrumentalization" (personal communication with Beauchamp, January 25, 2006). Beauchamp & Childress believe that the important content of human dignity is adequately captured in their four principles and associated rules, which they believe are clearer than the concept of human dignity (personal communication with Beauchamp, January 25, 2006).

Regarding the principle of vulnerability of Rendtorff & Kemp, one could argue that the vulnerability of the patient is a condition that should not be considered as an ethical principle. However, Rendtorff & Kemp define this principle in the phenomenological tradition and write: "The notion of vulnerability is especially appealing to the modern human being by the fact that in ordinary language it is not only descriptive but at the same time explicitly normative. Thus, in the discussion of vulnerability we cannot maintain a sharp distinction between fact and norm because vulnerability is mostly always already an ethical concept" (Rendtorff & Kemp, 2000, p. 46). Hence, Rendtorff & Kemp believe that the fact that a person is vulnerable entails a demand to respect the vulnerability of that person. Beauchamp & Childress do include reflections on vulnerable persons in their theory. However, these reflections are part of their principle of justice. In the chapter on the principle of justice, they first go into problems of health care distribution, next they focus on problems about the vulnerability of human research subjects at risk of exploitation.

In the 1970s to 1990s, the notion of "vulnerable groups" was often used within bioethics, however, Beauchamp & Childress do not think that "vulnerable" is an appropriate concept covering a class of people, since some of these people might not be vulnerable in the relevant senses. It would be better to speak of vulnerabilities, hence to make a list indicating forms and conditions of vulnerability. For instance, all pregnant women are not by category vulnerable, but some pregnant women might be in some respects (Beauchamp & Childress, 2009, pp. 253-254). Regarding vulnerability, Beauchamp & Childress go specifically into the recruitment and enrolment in clinical research of the economically disadvantaged. These persons are often "disadvantaged by the social lottery" (Beauchamp & Childress, 2009, p. 253). By "economically disadvantaged" they mean "persons who are impoverished, may lack significant access to health care, may be homeless, may be malnourished, and so fourth", and yet "possess a basic competence to reason, deliberate, decide, and consent" (Beauchamp & Childress, 2009, p. 253). Beauchamp & Childress think it would be unjust to exclude economically disadvantaged persons categorically because this would deprive them of the freedom to choose, and it would often be harmful to their financial interests (since if they were enrolled in research, their participation would be honoured by payment) (Beauchamp & Childress, 2009, p. 254). As can be seen, Beauchamp & Childress' analysis of the vulnerable economically disadvantaged only concerns persons who "possess a basic competence to reason, deliberate, decide, and consent" (Beauchamp & Childress, 2009, p. 253). This can be seen in contrast to Rendtorff & Kemp, who wish to protect persons who are not able to reason by invoking the principles of dignity, integrity, and vulnerability. Based on the different philosophical traditions from which they develop their ethical theories, Beauchamp & Childress take their point of departure in a stronger human subject than Rendtorff & Kemp do. For Rendtorff & Kemp, the human person is by definition vulnerable and fragile. In Beauchamp & Childress' theory, if the patient is constrained or vulnerable, the principles of beneficence and nonmaleficence protect the patient. For Beauchamp & Childress, vulnerability can be basically conceived as a foundation of the principle of nonmaleficence as they write: (since some persons) "may be vulnerable in several ways to influences that introduce a significant risk of *harm*[4]. Their situation may leave them lacking in critical

[4] Emphasis by the author of this article.

resources and forms of social powers that might have been created on their behalf. Hence, they may not be able to resist or refuse acceptance of the risk involved, requiring trade-offs among their interests" (Beauchamp & Childress, 2009, p. 254).

Some of the moral dimensions of the principle of integrity of Rentorff & Kemp are included in the principle of respect for autonomy of Beauchamp & Childress. Rendtorff & Kemp's principle of integrity emphasises the following understanding of integrity: "integrity as a personal sphere for experiences, creativity, and personal self-determination (autonomy)" (Rendtorff & Kemp, 2000, p. 40). This is somewhat comparable to the following part of the principle of respect for autonomy by Beauchamp & Childress: "to respect autonomous agents is to acknowledge their right to hold views, to make choices, and to take actions based on their personal values and beliefs" ... "It requires more than noninterference in others' personal affairs. It includes, in some contexts, building up or maintaining others' capacities for autonomous choice while helping to allay fears and other conditions that destroy or disrupt autonomous action" (Beauchamp & Childress, 2009, p. 103).

Besides respect for autonomy, Beauchamp & Childress' theory includes principles of beneficence, nonmaleficence, and justice. One could say that the principle of nonmaleficence is the foundation of Rendtorff & Kemps' four principles, since they developed their theory with the purpose of protecting the human being against biomedical interventions. They intend the principles to function as practical guidelines in order to use technology in an appropriate way so that it does not lead to the destruction of the vulnerable and weak (Rendtorff & Kemp, 2000, pp. 17-20). Rendtorff & Kemp write that the principles are integrated into a vision of social justice founded on mutual solidarity and responsibility (Rendtorff & Kemp, 2000, p. 56). Hence, they include reflections on justice. However, the theory of Rendtorff & Kemp does not include a positive obligation of beneficence such as Beauchamp & Childress' theory does. Beauchamp & Childress stress that we ought to prevent and remove evil or harm and that we ought to do and promote good (Beauchamp & Childress, 2009, p. 151). The theory of Rendtorff & Kemp focuses on the protection of the human subject, not on promoting good.

The basic difference between the theories of Beauchamp & Childress and Rendtorff & Kemp is that they include different views on human nature because of the different philosophical tradition from which they developed their theories. This difference in views on human nature is reflected in the choice of principles and their content. However, worth note is that Beauchamp & Childress' theory includes positive obligations to respect the autonomy of persons and positive obligations of beneficence to remove harm and promote good. Rendtorff & Kemp's theory on the other hand does not include positive obligations, which means that we do not have to take positive steps, but can remain passive. Including such positive obligations would still be in agreement with the philosophical tradition they base their theory on.

The theory of Rendtorff & Kemp is an example of a general critique of the framework of Beauchamp & Childress' theory. Generally speaking, the critique of Beauchamp & Childress' theory centres around two main objections: 1) Beauchamp & Childress focus too much on individualism, individual rights, and the dominance of the individual in the physician-patient relationship, and 2) Beauchamp & Childress have too narrow a focus on the self as independent and under rational control. The critics do not believe in an independent rational will that is negligent to emotions, communal life, and reciprocity. Recently, some feminists and care ethicists have tried to revise individualistic conceptions of autonomy by use of a model based on relational autonomy, focusing on the belief that persons are socially embedded and that agents' identities are formed within social relationships (Donchin, 1995; Thomasma, 1995; deMoissac, 1996; Rendtorff & Kemp, 2000; Beauchamp & Childress, 2001; Maeckelberghe, 2004). Beauchamp & Childress answer this critique by saying:

> "Many commentators have said that the principle of respect for autonomy overrides all other moral considerations in our work, reflecting a distinctly American bias weighting autonomy higher than other principles. This interpretation is profoundly mistaken. In a properly structured theory, respect for autonomy is not an excessively individualistic, absolutistic, or overriding notion that emphazises individual rights to the neglect or exclusion of social responsibilities. The principle of respect for autonomy is not so treated in this book and has not been so treated from the first edition to the present. We have always argued that competing moral considerations validly override this principle under many conditions" (Beauchamp & Childress, 2009, p. viii).

REFERENCES

Beauchamp TL (1995). Principlism and its alleged competitors. *Kennedy Inst Ethics J* 5(3):181-198.

Beauchamp TL (2000). Reply to Strong on principlism and casuistry. *J Med Philos* 25(3):342-347.

Beauchamp TL (2003). A defense of the common morality. Kennedy Inst Ethics J 13(3):259-74.

Beauchamp TL, Childress JF (1989). *Principles of biomedical ethics.* 3rd ed. Oxford: Oxford University Press.

Beauchamp TL, Childress JF (2001). *Principles of biomedical ethics.* 5th ed. Oxford: Oxford University Press.

Beauchamp TL, Childress JF (2009). *Principles of biomedical ethics.* 6th ed. Oxford: Oxford University Press.

DeGrazia D (1992). Moving forward in bioethical theory: theories, cases, and specified principlism. *J Med Philos* 17:511-539.

DeGrazia D, Beauchamp TL (2001). *Philosophy. In: Methods in medical ethics.* Sugarman J, Sulmasy DP (eds.), pp. 3-19. Washington D.C.: Georgetown University Press.

deMoissac DM, Warnock FF (1996). The evolution of caring within bioethics: provision for relationship and context. *Nurs Ethics* 3(3):190-201.

Donchin A (1995). Reworking autonomy: toward a feminist perspective. *Camb Q Healthc Ethics* 4(1):44-55.

Ebbesen M, Pedersen BD (2007a). Using empirical research to formulate normative ethical principles in biomedicine. *Med Health Care Philos* 10(1):33-48.

Ebbesen M, Pedersen BD (2007b). Empirical investigation of the ethical reasoning of physicians and molecular biologists – the importance of the four principles of biomedical ethics. *Philos Ethics Humanit Med* 2:23.

Ebbesen M, Pedersen BD (2008a). The principle of respect for autonomy - concordant with the experience of physicians and molecular biologists in their daily work? *BMC Med Ethics* 9:5.

Ebbesen M, Pedersen BD (2008b). The role of ethics in the daily work of oncology physicians and molecular biologists – results of an empirical study. *Bus Prof Ethics J.* Accepted for publication.

Holm S (1995). Not just autonomy - the principles of American biomedical ethics. *J Med Ethics* 21(6):332-338.

Kappel K (2006). The meta-justification of reflective equilibrium. *Ethical Theory Moral Pract* 9:131-47.

King N, Churchill L (2000). Ethical principles guiding research on child and adolescent subjects. *J Interpers Violence* 15(7):710-724.

Maeckelberghe E (2004). Feminist ethic of care: a third alternative approach. *Health Care Anal* 12(4):317-27.

Petersson B (1998). Wide reflective equilibrium and the justification of moral theory. *In: Reflective equilibrium.* Burg W van der, Willigenburg T van (eds.), pp. 127-134. The Netherlands: Kluwer Academic Publishers.

Rawls J (1971*). A theory of justice. 4th printing.* Cambridge, Mass. London: Belknap, 2001.

Rawls J (2001). *Justice as fairness. A restatement.* Cambridge, Mass. London: Belknap.

Rendtorff J (2002). Basic ethical principles in European bioethics and biolaw: autonomy, dignity, integrity and vulnerability – towards a foundation of bioethics and biolaw. *Med Health Care Philos* 5:235-244.

Rendtorff J, Kemp P (2000). Basic ethical principles in European bioethics and biolaw. Vol. 1: autonomy, dignity, integrity and vulnerability. *Denmark: Centre for ethics and law.*

Richardson HP (2000). Specifying, balancing, and interpreting bioethical principles. *J Med Philos* 25(3):285-307.

Ross WD (1930). *The right and the good.* Oxford: The Clarendon Press.

Strong C (2000). Specified principlism: what is it, and does it really resolve cases better than casuistry? *J Med Philos* 25(3):323-341.

Thomasma DC (1995). Beyond autonomy to the person coping with illness. *Camb Q Healthc Ethics* 4(1):12-22.

In: Bioethics: Issues and Dilemmas
Editor: Tyler N. Pace
ISBN: 978-1-61728-290-4

Chapter 6

OOPHORECTOMY SPECIMENS AS A POTENTIAL SOURCE OF OOCYTES FOR HUMAN EMBRYONIC STEM CELL RESEARCH

Mikyung Kim* and Kyu Won Jung≠
[1]KAIST , Graduate School of Innovation & Technology Management, Goosung-dong, Yooseong-gu, DaeJeon, South Korea
[2]Law School 308, Hanyang University, Seoul, 791, South Korea

ABSTRACT

Human embryonic stem cell research aimed at producing patient-specific tissues for cell-based therapies needs to create embryos through somatic cell nuclear transfer (SCNT), for which human oocytes will be necessary. However, oocyte donation is a risky and uncomfortable process, partly because of hormonal stimulation to induce ovulation. *In vitro* maturation (IVM) of immature human oocytes is in clinical use for infertility treatments. This article discusses the feasibility of, and ethical

* Mikyung Kim, MD, PhD, JD , KAIST , Graduate School of Innovation & Technology Management, 335 Gwahang-no, Bldg E7/Rm 2102, Goosung-dong, Yooseong-gu, DaeJeon 305-701, South Korea, phone: +82-42-350-4237, FAX: +82-42-350-4240, mikyung.kim@kaist.ac.kr

≠ Kyu Won Jung, MD, PhD, Law School 308, Hanyang University, 17 Haengdang-dong, Seongdong-gu, Seoul, 133-791, South Korea, Phone Number +82-2-2220-1309, dike1@hanyang.ac.kr

and legal considerations for, using IVM of human oocytes retrieved from surgically removed ovaries for SCNT; and concludes IVM of human oocytes from surgically removed ovaries clearly warrants exploration.

INTRODUCTION

Much of the research into human embryonic stem cells has been performed with cell lines derived from previously created embryos, usually those left over from attempted *in vitro* fertilization. However, for human embryonic stem cell research aimed at producing patient-specific tissues for cell-based therapies, it is necessary to create embryos through somatic cell nuclear transfer (SCNT), for which oocytes, presumably human, will be necessary. Although a recently developed induced pluripotent stem cell research technique employing skin fibroblasts appears to reduce the necessity of nuclear transfer, it does not replace the need of embryonic stem cell research using nuclear transfer owing to technical hurdles that must still be overcome. For example, researchers must determine a means to prevent carcinogenesis by the viral vectors used in the induced pluripotent stem cells before they can be used in humans.

Several laboratories worldwide are currently using this technique in human stem cell research,[1] including the Harvard University Stem Cell Institute, the University of California San Francisco, and the University of California Los Angeles.[2] The British Human Fertilisation and Embryology Authority granted a license to use nuclear transfer to research groups at the Newcastle Center for Life[3] and the Roslin Institute[4]. In April 2009, after a three-year moratorium on human stem cell research, South Korean officials approved a new SCNT project involving human eggs.[5]

1 Elizabeth Weise, "Cloning race is on again," USA Today (January 17, 2006, retrieved July 10, 2009) at http://www.usatoday.com/tech/science/genetics/2006-01-17-stem-cell-rejuvenated_x.htm

2 Id. Accessed on July 10, 2009 at http://www.usatoday.com/tech/science/genetics/2006-01-17-stem-cell-rejuvenated_x.htm#chart

3 Stephen Pincock, "Newcastle centre gains licence for therapeutic cloning," BMJ 2004;329:417 (21 August), doi:10.1136/bmj.329.7463.417 Accessed on July 10, 2009at http://www.bmj.com/cgi/content/full/329/7463/417?etoc

4 "Dolly scientists' human clone bid," BBC News (September 28, 2004, retrieved July 10, 2009) at http://news.bbc.co.uk/2/hi/health/3695186.stm

5 S. Korea conditionally lifts ban on stem cell research. 2009/04/29 15:32 KST. http://english.yonhapnews.co.kr/news/2009/04/29/0200000000AEN20090429007100320.HTML

Oocyte donation is a risky and uncomfortable process. It is fraught with both practical and ethical difficulties, as the deceptions about the numbers and sources of human oocytes used in Woo Suk Hwang's work reveal.[6] The main discomfort and risk in oocyte donation results from hormonal stimulation to induce ovulation of ten or twenty mature oocytes per menstrual cycle. Because of such ethical and health implications, the 2009 National Institutes of Health Guidelines on Human Stem Cell Research, implementing new U.S. policy, still do not allow for funding of research using hESCs derived from embryo-like entities created by SCNT.[7]

In vitro maturation (IVM) of oocytes is in clinical use for infertility treatments when hormonal stimulation cannot be used. The Newcastle group already attempted to use this technique to obtain human oocytes for embryonic stem cell research. This article explores the feasibility of IVM of immature human oocytes to provide a source of human oocytes for stem cell research, and thereby avoid the medical and ethical problems of standard egg donation. Immature human oocytes can be taken either directly from female volunteers or from ovaries that have been surgically removed for other reasons.

First, the present article describes the feasibility of using the retrieval and *in vitro* maturation of immature human oocytes taken from surgically removed ovaries for SCNT. Second, it proposes a model consenting process for oocyte procurement from surgically removed ovaries for SCNT. Third, the article discusses potential advantages in oocyte procurement from surgically removed ovaries, and concludes that oocyte retrieval and *in vitro* maturation of human oocytes from surgically removed ovaries clearly warrants exploration.

6 Magnus D, Cho MK. Issues in oocyte donation for stem cell research. Science 2005;5729(308):1747-8; Magnus D, Cho MK. A commentary on oocyte donation for stem cell research in South Korea. Am J Bioeth 2006 Jan-Feb;1(6):W23-4; CIRM guidelines Accessed on July 10, 2009 at http://www.cirm.ca.gov/workgroups/pdf/oocyte_guidelines.pdf ; Greely HT (2006), Moving Human Embryonic Stem Cells from Legislature to Lab: Remaining Legal and Ethical Questions. PLoS Med 3(5): e143. doi:10.1371/journal.pmed.0030143 Published: February 28, 2006

7 Raynard S. Kington. National Institutes of Health Guidelines on Human Stem Cell Research. http://stemcells.nih.gov/policy/2009guidelines.htm

I. The Feasibility of Using The Retrieval and *In Vitro* Maturation of Immature Human Oocytes Taken from Surgically Removed Ovaries for Somatic Cell Nuclear Transfer

A. Retrieval and *In Vitro* Maturation of Immature Human Oocytes for Clinical Use

Immature human oocyte retrieval followed by *in vitro* maturation (IVM) has progressed from a research level into clinical practice in a short time [the last several decades], producing live births consistently.[8] The technology was first proven effective in *in vitro* fertilization. Table 1 summarizes the results of major reports using *in vitro* fertilization combined with *in vitro* maturation. Its usage has currently expanded into preserving fertility in oncology, as well as into pre-implantation diagnosis.

1. Development of In Vitro Fertilization Combined with In Vitro Maturation

The first IVM of human oocytes and the first fertilization of an *in vitro* matured oocyte were reported by Edwards (1965) [9] and Edwards et al. (1969), respectively.[10] The first birth was achieved by Veeck et al. (1983) who obtained immature oocytes during a cycle with ovarian stimulation. [11] Subsequently, other researchers have reported cases of live births following IVM of oocytes obtained from non-stimulated ovaries.[12 13 14] Cha et al. (1991)

---

8 Söderström-Anttila V, Mäkinen S, Tuuri T, Suikkari A-M (2005) Favourable pregnancy results with insemination of in vitro matured oocytes from unstimulated patients Hum. Reprod. Advance Access published on June 1, 2005, DOI 10.1093/humrep/deh768. Hum. Reprod. 20: 1534-1540 (1540).

9 Edwards R (1965) Maturation in vitro of mouse, sheep, cow, pig, rhesus monkey and human ovarian oocytes. Nature 208, 349-51.

10 Edwards R, Bavister B, Steptow P (1969) Early stages of fertilization in vitro of human oocytes matured in vitro. Nature 221, 632-35.

11 Veeck L, Worthman J, Witmyer J, Sandow B, Acosta A, Garcia J, Jones G, Jones H (1983) Maturation and fertilization of morphologically immature human oocytes in a program of in vitro fertilization. Fertil Steril 39, 594-602.

12 Cha KY, Koo JJ, Ko JJ, Choi DH, Han SY, Yoon TK. Pregnancy after in vitro fertilization of human follicular oocytes collected from nonstimulated cycles, their culture in vitro and their transfer in a donor oocyte program. Fertil Steril 1991;55:109-13.

13 Trounson AO, Wood C, Kausche A. In vitro maturation and the fertilization and developmental competence of oocytes recovered from untreated polycystic ovarian patients. Fertil Steril 1994;62:353-62.

described the successful IVM of immature oocytes aspirated from ovarian biopsy specimens taken at the time of cesarean section, resulting in a triplet birth.[15] Trounson et al. (1994) reported the first pregnancy achieved in patients with polycystic ovary syndrome (PCOS) from immature oocytes recovered by puncture during non-stimulated cycles.[16] The cryopreservation of embryos using in-vitro matured oocytes has resulted in normal pregnancies.[17] Thus far, the delivery of more than 300 babies has been reported after IVM followed by assisted reproduction treatment.[18] Researchers also reported a case where they successfully generated a sufficient number of embryos from natural-cycle IVF combined with an IVM procedure for pre-implantation genetic diagnosis.[19]

2. Methods of Retrieval and In Vitro Maturation of Immature Oocytes for In Vitro Fertilization

For infertility treatment, researchers have developed various methodologies for immature oocyte retrieval and IVM, which were modified from the methods originally described by Trounson et al. in 1994.[20] However, none of them uses controlled ovarian stimulation with fertility drugs or hormones [gonadotropins] for immature oocyte recovery.[21]

---

14 Barnes FL, Crombie A, Gardner DK, Kausche A, Lacham-Kaplan O, Suikkari A-M, Tiglias J, Wood C, Trounson AO. Blastocyst development and birth after in-vitro maturation of human primary oocytes, intracytoplasmic sperm injection and assisted hatching. Human Reproduction 1995;10:3243-47.

15 Cha KY, Koo JJ, Ko JJ, Choi DH, Han SY, Yoon TK. Pregnancy after in vitro fertilization of human follicular oocytes collected from nonstimulated cycles, their culture in vitro and their transfer in a donor oocyte program. Fertil Steril 1991;55:109-13.

16 Trounson AO, Wood C, Kausche A. Fertil Steril 1994;62:353-62.

17 Suikkari A-M, Tulppala M, Tuuri T, Hovatta O, Barnes F. Luteal phase start of low-dose FSH priming of follicles results in an efficient recovery, maturation and fertilization of immature human oocytes. Hum. Reprod. 15: 747-751.; Chian, R.C., Gulekli, B., Buckett, W.M. and Tan, S.L. (2001) Pregnancy and delivery after cryopreservation of zygotes produced by in-vitro matured oocytes retrieved from a woman with polycystic ovarian syndrome. Hum. Reprod., 16, 1700–1702.

18 Le Du A, Kadoch IJ, Bourcigaux N, Doumerc S, Bourrier M-C, Chevalier N, Fanchin R, Chian R-C, Tachdjian G, Frydman R, Frydman N. In vitro oocyte maturation for the treatment of infertility associated with polycystic ovarian syndrome: the French experience. Hum. Reprod. Advance Access published on February 1, 2005, DOI 10.1093/humrep/deh603. Hum. Reprod. 20: 420-424. (quoting Mikkelsen A (2004) In vitro maturation of human ova. Int Congr Ser 1266, 160–166) .

19 Ao A, Jin S, Rao D, Son W-Y, Chian R-C, Tan SL. First successful pregnancy outcome after preimplantation genetic diagnosis for anueploidy screening in embryos generated from natural-cycle in vitro fertilization combined with an in vitro maturation procedure. Fertil Steril 2006;85:1510.e9-11.

20 Trounson AO, Wood C, Kausche A. Fertil Steril 1994;62:353-62.

21 Barnes FL, et al. Human Reproduction 1995;10:3243-47, p3244

Oocytes were recovered by ultrasound-guided transvaginal aspiration of all visible follicles (between 2 and 10 mm in diameter) except the dominant follicle[22], on days 10-12 of a spontaneous menstrual cycle, or when the size of a leading follicle reached 10 mm and the endometrium had a thickness of at least 5 mm.[23] A specially designed single-lumen aspiration needle[24] and low or reduced aspiration pressure[25] were used for oocyte collection.[26] [27] During the collection, patients received an IV sedation,[28] which may have been combined with local anesthesia[29]. [30]

Some researchers have carried out IVM in natural cycles without any hormonal priming. Other researchers have used pretreatment with a follicle-stimulating hormone (FSH) [31] and/or human chorionic gonadotropin (hCG) (10,000IU; subcutaneous or intramuscular injection), 36 hours before immature oocyte retrieval,[32] or human menopausal gonadotropin[33].

---

22 Some researchers aspirated the dominant follicle and inseminated all oocytes from the dominant follicle 4-6 hours after aspiration. In conventional IVF, aspiration is performed on follicles at least 18-20 mm in diameter after ovarian stimulation.

23 Barnes FL, et al. Human Reproduction 1995;10:3243-47

24 A 17, 19, or 21 gauge-needle was used for aspiration.

25 The aspiration pressure ranged 75-100 mm Hg.

26 Trounson AO, Wood C, Kausche A. Fertil Steril 1994;62:353-62.

27 On day 3 of the patient's spontaneous menstrual cycle, a baseline transvaginal ultrasound scan was performed to observe small follicles (less than 4 mm in diameter) developed in both ovaries. On day 8 of her cycle, ultrasound scan was performed to observe whether a dominant follicle (13 mm in diameter) had developed.

28 Sedation was performed by an intravenous administration of 2 mg midazolam and 175 μg fentanyl.

29 Paracervical block with 10 mL of 1% lidocaine was performed for local anesthesia.

30 Child TJ, Abdul-Jalil AK, Gulekli B, Tan SL. In vitro maturation and fertilization of oocytes from unstimulated normal ovaries, polycystic ovaries, and women with polycystic ovary syndrome. Fertil Steril 2001;76:936-42 (p941) (The authors speculated that oocyte aspiration for IVM involving multiple punctures would take longer and be more painful than oocyte retrieval for conventional IVF.)

31 Wynn, P., Picton, H.M., Krapez, J. et al. (1998). Randomized study of oocytes matured after collection from unstimulated or mildly stimulated patients. Hum. Reprod., 13, 3132–3138; Mikkelsen AL, Smith SD, Lindenberg S. In-vitro maturation of human oocytes from regularly menstruating women may be successful without follicle stimulating hormone priming. Hum. Reprod. 14: 1847-1851; Suikkari A-M, Tulppala M, Tuuri T, Hovatta O, Barnes F. Luteal phase start of low-dose FSH priming of follicles results in an efficient recovery, maturation and fertilization of immature human oocytes. Hum. Reprod., Apr 2000; 15: 747 – 751; Mikkelsen AL, Lindenberg S. Morphology of in-vitro matured oocytes: impact on fertility potential and embryo quality. Human Reproduction 16(8):1714-18, 2001.

32 Chian RC, Buckett WM, Tulandi T, Tan SL. Prospective randomized study of human chorionic gonadotrophin priming before immature oocyte retrieval from unstimulated women with polycystic ovarian syndrome. Hum. Reprod., Jan 2000; 15: 165-70; Child TJ, Abdul-Jalil AK, Gulekli B, Tan SL. In vitro maturation and fertilization of oocytes from unstimulated normal ovaries, polycystic ovaries, and women with polycystic ovary

Follicular aspirates were collected in culture tubes, washed through a filter to recover and isolate oocytes, and rinsed to remove erythrocytes. Cumulus-oocyte complexes were identified under a stereomicroscope. Immature oocytes, which were determined by cumulus compaction and/or the presence or absence of a germinal vesicle in their cytoplasm or the first polar body extruded into the perivitelline space, were cultured in the maturation medium or tissue culture medium supplemented with FSH, luteinizing hormone (LH), and/or hCG. After incubating for 24 and 48 hours, the maturity of the oocytes was determined by microscopic examination. Mature oocytes, which were denuded of cumulus cells, were considered to be ready for fertilization.[34]

3. Expansion of the Indication of In Vitro Fertilization Combined with In Vitro Maturation Treatment

Initially, women with polycystic ovary syndrome (PCOS) were considered suitable candidates for the IVM procedure. Women with PCOS can provide a good yield of immature oocytes, capable of further development.[35] Moreover, those women are at high risk of ovarian hyper-stimulation syndrome (OHSS) in the conventional IVF procedure.[36] [37] However, the indication for the IVM

---

syndrome. Fertil Steril 2001;76:936-42 (The authors demonstrated that when using hCG priming before oocyte retrieval, immature oocytes retrieved from normal ovaries could achieve final maturation rates comparable to oocytes retrieved from women with PCOS); Child TJ, Sylvestre C, Pirwany I, Tan SL. Basal serum levels of FSH and estradiol in ovulatory and anovulatory women undergoing treatment by in-vitro maturation of immature oocytes. Human Reproduction 17(8):1997-2002 (2002); Lin Y-H, Hwang J-L, Huang L-W, Mu S-C, Seow K-M, Chung J, Hsieh B-C, Huang S-C, Chen C-Y, Chen P-H. Combination of FSH priming and hCG priming for in-vitro maturation of human oocytes. Hum. Reprod., 2003; 18: 1632-6.

33 Jaroudi KA, Hollanders JMG, Elnour AM, Roca GL, Atare AM, Coskun S. Embryo development and pregnancies from in-vitro matured and fertilized human oocytes, Human Reproduction 14(7):1749-51, 1999 (The patients were subjected to limited exposure to human menopausal gonadotrophin and no exposure to HCG.).

34 Chian RC, Buckett WM, Tulandi T, Tan SL. Prospective randomized study of human chorionic gonadotrophin priming before immature oocyte retrieval from unstimulated women with polycystic ovarian syndrome. Hum. Reprod., Jan 2000; 15: 165 – 170 (p166).

35 Women with PCO or PCOS are known to have numerous antral follicles within their ovaries. Cavilla JL, Kennedy CR, Baltsen M, Klentzeris LD, Byskov AG, Hartshorne GM. The effects of meiosis activating sterol in in-vitro maturation and fertilization of human oocytes from stimulated and unstimulated ovaries. Hum Repro 2001;16:547-55 (pp 548). (A classical polycystic ovary has several antral follicles up to 10 mm diameter around the periphery of the ovary. Although these follicles are not actively growing, they contain immature oocytes that may be capable of further development.)

36 Compared to conventional IVF, the greatest advantage of IVM in women with PCO or PCOS is to avoid their increased risk of developing ovarian hyper-stimulation syndrome under stimulation with exogenous gonadotropins.

treatment soon expanded to women with polycystic ovaries (PCO) and to women with normal ovaries. Currently, natural cycle IVF combined with IVM treatment is suggested as an alternative treatment for women with various causes of infertility, including women with diminished ovarian response.[38 39 40]

4. The Outcome of In Vitro Fertilization Combined with In Vitro Maturation

According to a literature review by Le Du et al. (2005),[41] the average number of immature oocytes collected by puncture from patients with PCOS ranged from 10.3 to 23.1, the maturation rate from 44% to 84%, and the fertilization rate from 26% to 95%, respectively. In a 2004 report, it was stated that the clinical pregnancy and implantation rates of IVM treatment in infertile women with ovulatory PCO-like ovaries or anovulatory PCOS had reached approximately 30%-35% and 10%-15%, respectively.[42] In particular, about 100 births have been reported following IVM of oocytes derived from non-stimulated cycles in women with PCOS.[43]

Fewer immature oocytes have been retrieved from normal ovaries than from polycystic ovaries.[44] Table 1 reveals the average number of immature

37 Chian RC, Lim JH, Tan SL, State of the art in in vitro oocyte maturation. Curr Opin Obstet Gynecol 2004;16:211-19 (pp214).

38 Chian R-C, Buckett WM, Jalil AKA, Son W-Y, Sylvestre C, Rao D, Tan SL. Natural-cycle in vitro fertilization combined with in vitro maturation of immature oocytes is a potential approach in infertility treatment, Fertil Steril 2004;82:1675-8. ("It is appropriate to modify the current protocol of IVM treatment for women with ovulatory PCO-like ovaries or anovulatory PCOS and avoid canceling cycles if a dominant follicle develops.")

39 Friden B, Hreinsson J, Hovatta O. Birth of a healthy infant after in vitro oocyte maturation and ICSI in a woman with diminished ovarian response: Case report. Hum Rep 2005;20:2556-8.

40 Liu J, Lu G, Qian Y, Mao Y, Ding W. Pregnancies and births achieved from in vitro matured oocytes retrieved from poor responders undergoing stimulation in in vitro fertilization cycles. 2003;80:447-49.

41 Le Du A, Kadoch IJ, Bourcigaux N, Doumerc S, Bourrier MC, Chevalier N, Fanchin R, Chian R-C, Tachdjian G, Frydman R, Frydman N. In vitro oocyte maturation for the treatment of infertility associated with polycystic ovarian syndrome: the French experience. 2005;20:420-24. Table II. on p423.

42 Chian RC, Buckett WM, Tan SL. In-vitro maturation of human oocytes. Reprod Biomed Online 2004; 8:148-66.

43 Le Du A, Kadoch IJ, Bourcigaux N, Doumerc S, Bourrier M-C, Chevalier N, Fanchin R, Chian R-C, Tachdjian G, Frydman R, Frydman N. In vitro oocyte maturation for the treatment of infertility associated with polycystic ovarian syndrome: the French experience. Human Reproduction. 20(2):420-424, February 2005.

44 Child TJ, Abdul-Jalil AK, Gulekli B, Tan SL. In vitro maturation and fertilization of oocytes from unstimulated normal ovaries, polycystic ovaries, and women with polycystic ovary syndrome. Fertil Steril 2001;76:936-42. Table 1 on page 938 (reported that significantly

oocytes procured from women with a presumed regular ovulatory cycle or normal ovaries was 0.7 to 11.2 per aspiration. However, once retrieved, immature oocytes from normal and polycystic ovaries did not show any qualitative differences. The oocytes appear to retain similar developmental potential in terms of their final maturation rates.[45 46] The oocytes exhibited no significant differences in morphology, which is known to correlate with embryo quality and cleavage rate.[47] However, the fewer cleaving embryos generated per cycle restricted choices when selecting embryos for transfer,[48] leading to lower rates of implantation and pregnancy in women with normal ovaries compared to those with polycystic ovaries.[49 50]

Fewer live births per cycle in IVF combined with IVM treatment than in the conventional IVF treatment has been attributed to factors related to the oocytes and the endometrium, respectively.[51] First, incomplete cytoplasmic maturation of an oocyte would lead to asynchronized nuclear-cytoplasmic

fewer oocytes were retrieved from women with normal ovaries (5.1+3.7) compared to women with PCO (10.0+5.1) or PCOS (11.3+9.0)).

45 Child TJ, Abdul-Jalil AK, Gulekli B, Tan SL. In vitro maturation and fertilization of oocytes from unstimulated normal ovaries, polycystic ovaries, and women with polycystic ovary syndrome. Fertil Steril 2001;76:936-42. (The rates of oocyte maturation, fertilization, and embryo cleavage as well as the proportion of atretic oocytes retrieved were comparable between the two groups of women.)

46 Mikkelsen AL, Andersson A-M, Skakkebæk NE, Lindenberg S. Basal concentrations of oestradiol may predict the outcome of in-vitro maturation in regularly menstruating women. 2001;16:862-67. pp866 (also found that the developmental potential did not significantly differ between the two groups).

47 Mikkelsen AL, Lindenberg S. Morphology of in-vitro matured oocytes: impact on fertility potential and embryo quality. Human Reproduction 16(8):1714-18, 2001.

48 Child TJ, Abdul-Jalil AK, Gulekli B, Tan SL. In vitro maturation and fertilization of oocytes from unstimulated normal ovaries, polycystic ovaries, and women with polycystic ovary syndrome. Fertil Steril 2001;76:936-42, 940.

49 Child TJ, Abdul-Jalil AK, Gulekli B, Tan SL. In vitro maturation and fertilization of oocytes from unstimulated normal ovaries, polycystic ovaries, and women with polycystic ovary syndrome. Fertil Steril 2001;76:936-42. The rates of pregnancy per transfer, implantation, and live birth were significantly lower in women with normal ovaries (1.5%, 4.0%, and 2.0%, respectively) than in women with PCO (8.9%, 23.1%, and 17.3%, respectively) or PCOS (9.6%, 29.9%, and 14.9%, respectively).

50 Mikkelsen AL, Smith S, Lindenberg S. Impact of oestradiol and inhibin A concentrations on pregnancy rate in in-vitro oocyte maturation. Hum. Reprod. 2000;15: 1685-90. (demonstrated that relatively favorable pregnancy results can be achieved in women with regular cycles: they obtained oocytes from women with normal unstimulated ovaries after a leading follicle of 10 mm diameter and an endometrium of at least 5 mm thickness were observed via ultrasound, and achieved a 77% fertilization rate, 87% embryo cleavage rate, 12.6% (per aspiration) and 17.4% (per transfer) pregnancy rates; and 8.8% implantation rate; nine healthy children were born, and two pregnancies underwent abortion.)

51 Requena A, Neuspiller F, Cobo AC, Aragones M, Garcia-Velasco JA, Remohi J, Simon C, Pellicer A. Endocrinological and ultrasonographic variations after immature oocyte retrieval in a natural cycle. Hum Reprod 2001;16:1833-37. 1833

maturation of the oocyte, and fertilization of the oocyte would generate poor embryo quality, even when the cleavage rate of the oocyte is in an acceptable range.[52] Such poor embryo quality could lead to an unsuccessful full term pregnancy.[53] Second, inadequate endometrial priming, which may occur in IVF combined with IVM treatment, tends to lead to implantation failure.[54] The two year follow-up data of 49 children born after *in vitro* maturation of human oocytes revealed that the obstetric and perinatal outcome was good. Thus, while the endometrial factor would not be relevant to SCNT that does not involve implantation, the lower efficiency for immature oocytes with respect to attaining full cytoplasmic maturation *in vitro* would be problematic in SCNT using immature oocytes, most likely in a quantitative manner.

B. Retrieval and *In Vitro* Maturation of Immature Human Oocytes Taken from Surgically Removed Ovaries for Somatic Cell Nuclear Transfer

1. Surgically Removed Ovaries as a Potential Source of Oocytes

Oophorectomy was among the top 10 procedures for women admitted to hospital for non-obstetric reasons in the United States, and totals of 485,000 and 458,000 female patients underwent oophorectomy in 2000 and 2003, respectively.[55] In particular, oophorectomy was the second most common procedure for female patients ages 18-44, and totals of 195,000 and 179,000 underwent this procedure in 2000 and 2003, respectively.[56] The number of ambulatory oophorectomies, unilateral and bilateral, and other operations on the ovaries, which primarily include surgeries such as a laparoscopic removal

52 Requena A, Neuspiller F, Cobo AC, Aragones M, Garcia-Velasco JA, Remohi J, Simon C, Pellicer A. Endocrinological and ultrasonographic variations after immature oocyte retrieval in a natural cycle. Hum Reprod 2001;16:1833-37. 1833

53 Mikkelsen AL, Andersson A-M, Skakkebæk NE, Lindenberg S. Basal concentrations of oestradiol may predict the outcome of in-vitro maturation in regularly menstruating women. 2001;16:862-67. 865.

54 Requena A, Neuspiller F, Cobo AC, Aragones M, Garcia-Velasco JA, Remohi J, Simon C, Pellicer A. Endocrinological and ultrasonographic variations after immature oocyte retrieval in a natural cycle. Hum Reprod 2001;16:1833-37. 1833

55 H. Joanna Jiang, Anne Elixhauser, Joyce Nicholas, Claudia Steiner, Carolina Reyes, Arlene S. Bierman, Care of Women in U.S. Hospitals, 2000, HCUP Fact Book No. 3

56 H. Joanna Jiang, Anne Elixhauser, Joyce Nicholas, Claudia Steiner, Carolina Reyes, Arlene S. Bierman, Care of Women in U.S. Hospitals, 2000, HCUP Fact Book No. 3

of an ovary, in U.S. hospitals in 2003 was 26,000 (14.7%) and 41,000 (59.5%), respectively.[57]

More information is required to estimate how many oophorectomy specimens are available for oocyte retrieval. First, information regarding the conditions for which the oophorectomy was performed in the first place is needed. In the case of one of the major hospitals in Korea, in 2005, 68 out of 407 (16.7%) oophorectomies were performed because of benign conditions involving the uterus and/or adnexa, including uterine leimyoma, adenomyosis, lymphangioleiomyomatosis, benign tumors of a contralateral ovary (mucinous cystadenoma, mature cystic teratoma, fibroma, etc.), chronic salpingitis, endometriosis, and various benign ovarian cysts (cystic follicle, follicular cyst, hemorrhagic/cystic corpus luteum, polycystic ovaries).[58] Second, information pertaining to the average *number of ovaries* removed annually through all procedures as well as the proportion aged between 18 and 44 among patients who undergo ambulatory ovary removal is also required. Calculating conservatively, at least 75,000 mature oocytes will still be available for SCNT from all the oophorectomy specimens.[59]

The same factors might influence the oocyte yield from an oophorectomy specimen as well as that from an ovary *in situ*. Reportedly, those factors include the donor's age, day of the menstrual cycle, early follicular count, ovarian volume, maximum ovarian stromal blood flow velocity, baseline antral follicle count, etc.[60] [61] [62] [63] Thus, aspiration of the oophorectomy specimen would result in a higher yield of immature oocytes in patients with

---

57 http://www.ahrq.gov/data/hcup/factbk9/factbk9e.htm#y

58 2005 Data from Department of Pathology, Samsung Seoul Hospital kindly provided by Dr. GY Kwon.

59 Assumptions are: 1) 30 % of patients underwent ambulatory procedure were between age 18-44; 2) all oophorectomies were counted as unilateral oophorectomies; 3) one immature oocyte was procured from one ovary; 4) the lowest maturation rate of immature oocyte (37.8%) was used for the calculation.

60 Cha KY, Do BR, Chi HJ, Yoon TK, Choi DH, Koo JJ, Ko JJ. Viability of human follicular oocytes collected from unstimulated ovaries and matured and fertilized in vitro. Reproduction, Fertility and Development 1992;4: 695-701.

61 Cha KY, Do BR, Chi HJ, Yoon TK, Choi DH, Koo JJ, Ko JJ. Viability of human follicular oocytes collected from unstimulated ovaries and matured and fertilized in vitro. Reproduction, Fertility and Development 1992; 4: 695-701.

62 Child TJ, Abdul-Jalil AK, Gulekli B, Tan SL. In vitro maturation and fertilization of oocytes from unstimulated normal ovaries, polycystic ovaries, and women with polycystic ovary syndrome. Fertil Steril 2001;76:936-42, 940.

63 Child TJ, Gulekli B, Tan SL. Success during in-vitro maturation of oocyte treatment is dependent on the numbers of oocytes retrieved which are predicted by early follicular phase transvaginal ultrasound measurement of the antral follicle count and peak ovarian stromal blood flow velocity. Hum Reprod 2001;16 (Abstract Book 1), p41.

PCO/PCOS, who commonly present with more than 20 small to medium sized antral follicles in each ovary.[64] However, it remains speculative as to how much greater cumulative yield can be achieved by multiple aspirations on surgically removed ovaries. Trounson et al. (2001) reviewed the studies reporting the number of immature oocytes recovered from unstimulated oophorectomy specimens without hCG priming and the *in vitro* maturation rate of those immature oocytes (see Table 2).[65] Table 3 summarizes more recent reports.

Some of the findings demonstrate the flexibility of the origins of oocytes for maturation. Cha et al. (1991) aspirated 270 [157] immature oocytes from 23 ovaries removed for various gynecological indications.[66] It has been reported that oocytes recovered at the end of pregnancy can be matured. A total of 268 immature oocytes were thus aspirated from small follicles of 51 patients at the time of Caesarean section.[67] More recently, Revel et al. (2004) obtained 17 oocytes for IVM from ovaries removed by staging a laparotomy in a 43-year old woman with PCOS and endometrial carcinoma.[68] No significant difference in the number of recovered oocytes, or in the development of immature follicular oocytes, was demonstrated between cyclic and non-cyclic ovaries.[69]

It is believed that approximately 20 antral follicles are selected and continue to the preovulatory stages of development during each cycle.[70] Thus, the maximum number of oocytes that can be aspirated from an ovary would be set around 20. It is also believed that there is a relationship between follicular

---

64 Trounson A, Anderiesz C, Jones G. Maturation of human oocytes in vitro and their developmental competence. Reproduction 2001;121:51-75. 55.

65 Excerpted from Table 5 and Table 6 (pp60-3) from Trounson A, Anderiesz C, Jones G. Maturation of human oocytes in vitro and their developmental competence. Reproduction 2001;121:51-75.

66 Cha KY, Koo JJ, Ko JJ, Choi DH, Han SY, Yoon TK. Pregnancy after in vitro fertilization of human follicular oocytes collected from nonstimulated cycles, their culture in vitro and their transfer in a donor oocyte program. Fertility and Sterility1991;55;109–13.

67 Hwu Y-M, Lee RK-K, Chen C-P, Su J-T, Chen Y-W, Lin S-P. Development of hatching blastocysts from immature human oocytes following in-vitro maturation and fertilization using a co-culture system. Human Reproduction 1998;13:1916-21, p1918 ("The mean number of immature oocytes retrieved from each patient was 5.3 (ranging from 1 to 15).")

68 Revel A, Safran A, Benshushan A, Shushan A, Laufer N, Simon A. In vitro maturation and fertilization of oocytes from an intact ovary of a surgically treated patient with endometrial carcinoma: case report, 2004;19:1608-1611.

69 Cha KY, Do BR, Chi HJ, Yoon TK, Choi DH, Koo JJ, Ko JJ. Viability of human follicular oocytes collected from unstimulated ovaries and matured and fertilized in vitro. Reproduction, Fertility and Development 1992;4: 695-701.

70 Hillier SG. Current concepts of the role of FSH and LH in folliculogenesis. Hum Reprod 1994;9:188-191.

size in the human ovary and oocytes capable of resuming meiosis *in vitro*, and there is a minimum size requirement for follicles to acquire developmental competence.[71 72] Therefore, the number of oocytes retrieved, matured in-vitro, and developed into the stage where SCNT in ideal conditions can be attempted would be set at approximately the number of follicles that are inherently capable of resuming meiosis and full development *in vitro*.

2. How to Retrieve and Mature the Oocytes from Surgically Removed Ovaries

The following procedure can be proposed. The removed ovaries are immediately transferred in phosphate buffered saline at 37° C to the laboratory. Aspiration is performed using a 17 gauge needle on all visible (antral) follicles of the ovary, and follicular aspirates containing immature oocytes undergo the IVM process. Ovarian cortical tissue slices are then prepared from the remaining ovary, for immediate tissue culture to initiate growth and development of primordial and primary follicles, or for cryopreservation for later use.

3. Optimization of the IVM Method to Support the Efficient and Full Development of Oocytes

Before applying *in-vitro* matured oocytes in SCNT, the IVM technique needs to be optimized to achieve efficient and complete maturation of immature oocytes under the technologies currently available. Where the potential of an embryo largely reflects the competence of the oocyte,[73] IVM currently undertaken does not appear to fully support the correct development of oocyte competence.[74 75] Researchers need to optimize their IVM method,

71 Tsuji K, Sowa M, Nakano R. Relationship between human oocyte maturation and different follicular sizes. Biology of Reproduction 1985;32; 413–7.

72 Trounson A, Anderiesz C, Jones G. Reproduction 2001;121:51-75. 52-53. The minimum size of follicle required for developmental competence in humans is 5-7 mm in diameter.

73 Cavilla JL, Kennedy CR, Baltsen M, Klentzeris LD, Byskov AG, Hartshorne GM. The effects of meiosis activating sterol on in-vitro maturation and fertilization of human oocytes from stimulated and unstimulated ovaries. Hum. Reprod. 2001;16: 547-555, p548.

74 Krisher RL. The effect of oocyte quality on development, J. Anim. Sci. 82 (2004) (E-Suppl.), pp. E14–E23, E14.

75 Researchers observed reduced developmental competence, particularly cleavage and development beyond the second cleavage stage (3-4 cells) (Trounson et al 1994; Cavilla et al. 2001), and anomalies in nuclear and cytoplasmic maturation, consistent with cell cycle deficiencies that would compromise both the meiotic and developmental competencies of in-vitro matured human germinal vesicle stage oocytes (Combelles CMH, Cekleniak NA, Racowsky C, Albertini DF. Assessment of nuclear and cytoplasmic maturation in in-vitro matured human oocytes. Hum. Reprod. 2002;17: 1006-1016.) Combelles et al. (2002) also

including assessment and adjustment of the conditions for different groups of subjects as well as for different developmental stages of aspirated oocytes used for IVM.

In summary, it is technologically feasible to employ the retrieval and *in vitro* maturation of immature human oocytes taken from surgically removed ovaries for SCNT.

II. Ethical and Legal Considerations

A. Rationale for Employing Oophorectomy Specimens in Human Embryonic Stem Cell Research

There are several types of human stem cell research.[76] Each has its own merits and demerits and we don't have enough information to know which one is better or will become a treatment method. The human embryonic stem cell is one of the potential candidates to treat diseases. Also, we may get much knowledge about the human developmental mechanism and life phenomenon and mechanism of diseases through human embryonic stem cell research.

The core problem in human embryonic stem cell research using SCNT is how to obtain human oocytes. First of all, there are few who are willing to donate oocytes. Second, even if a woman wants to donate her oocytes, she may not be allowed to donate because of her health conditions. A woman who is willing to donate her oocytes will undergo stimulation of the ovaries and this procedure may cause health problems. Researchers have been interested in retrieving immature oocytes in the absence of or with limited gonadotropin exposure, when intending to mature the oocytes in vitro for embryo transfer purposes, because gonadotropin stimulation is associated with some disadvantages in terms of costs and side effects, which include ovarian hyperstimulation, deep vein thrombosis, and other possible long-term effects on ovarian carcinogenesis.[77] If we use oocytes from surgically removed ovaries, patients need not undergo ovarian stimulation and are therefore not at

observed only the germinal vesicle stage oocytes with a distinct chromatin pattern, which was the predominant type observed following retrieval, were competent to complete meiotic progression to metaphase II in vitro.

76 For details, see Kyu Won Jung, Perspectives on Human Stem Cell Research, J. Cell. Physiol. 220: 535-537, 2009.

77 Jaroudi KA, Hollanders JMG, Elnour AM, Roca GL, Atare AM, Coskun S. Embryo Development and Pregnancies from In-vitro Matured and Fertilized Human Oocytes, Human Reproduction 14(7):1749-51, 1999.

risk for developing ovarian hyperstimulation syndrome and other undesired consequences involving fertility drugs.

Surgically removed ovaries will be discarded unless used for other purposes. Using theses ovaries for human embryonic stem cell research may not harm women's bodies or autonomies if there are proper informed consent procedures. Like using residual oocytes that are obtained during assisted fertilization, it can be justified legally or ethically.

Another main legal issue for human embryonic stem cell research is whether or not it can be protected under the freedom of scientific research. There have been hot debates about which U. S. Constitutional provision provides the freedom of scientific research. Some authors have suggested due process,[78] or equal protection.[79] However, the Free Speech Clause of the First Amendment has attracted the most attention related to the constitutional basis of freedom of scientific research.[80] In Korea, the freedom of research is clearly stipulated in Constitutional Law §22.[81]

Freedom of thinking and research is essential for human intellectual advances including scientific and academic advances. Most of the new scientific discoveries, such as the heliocentric theory and Darwinism, as well as new treatment methods such as assisted fertilization and organ transplantation originated from the new ways of thinking and the perception of facts. From this point of view, the freedom of research and the freedom of academy should be protected widely as constitutional rights. However, research that is related to human bodies or lives, such as human embryonic stem cell research and genetic treatment has some different implications from

78 Richard Delgado, David R. Millen, God, Galileo, and Government: Toward Constitutional Protection for Scientific Inquiry, 53 Wash. L. Rev. 349(1978), at 392-398; Roger H. Taylor, The Fear of Drawing the Line at Cloning, 9 B.U.J. Sci. & Tech. L. 379(2003), at 389-390.

79 Richard Delgado, David R. Millen, supra note 3, at 399-402.

80 Dana Remus Irwin, Freedom of Thought: The First Amendment and the Scientific Method, 2005 Wis. L. Rev. 1479; James R. Ferguson, Scientific Inquiry and the First Amendment, 64 Cornell L. Rev. 639(1979); Michael Davidson, First Amendment Protection for Biomedical Research, 19 Ariz. L. Rev. 893(1977); Peter M. Brody, The First Amendment, Governmental Censorship, and Sponsored Research, 19 J. C. & U. L. 199(1992); Steve Keane, The Case Against Blanket First Amendment Protection of Scientific Research: Articulating a More Limited Scope of Protection, 59 Stan. L. Rev. 505. Also there is an argument that the right to research relies on the connection between research and protected interests in free speech and medical treatment. John A. Robertson, Embryo Culture and the "Culture of Life": Constitutional Issues in the Embryonic Stem Cell Debate, 2006 U. Chi. Legal F. 1. In the cloning debate, the right to pursue medical treatment and reproductive rights are also suggested. Russell Krobkin, Stem Cell Century, Yale University Press, 2007, at 79-91.

81 Constitutional Law Article 21 ① All citizens shall enjoy freedom of learning and the arts. ② The rights of authors, inventors, scientists, engineers and artists shall be protected by Act.

those that are not related to human bodies or lives. Human subjects are almost always needed in biomedical research and may be harmed during the procedures.[82] Also, the results of research will be developed into treatment methods and will influence human bodies or lives. So, the freedom of research that uses human subjects may be restricted in some cases.[83]

Under what conditions, how to restrict the freedom of scientific research, and the range of restrictions are very important and controversial issues. Research itself includes all activities such as thinking, reading and investigating or experimenting in pursuit of the truth. In principle, these activities that are performed only in the laboratory or office and have no direct influence on others should have unlimited protection under the Constitution. However, the publication of research results can be restricted in some cases such as when they would harm an innocent person.[84] Human embryonic stem cell research uses human oocytes and human somatic cells, so this research can be restricted in some cases. There are some Constitutional provisions or principles that can provide reasonable grounds to restrict human embryonic stem cell research. First of all, freedom of human embryonic stem cell research may be restricted based on human dignity.[85] However, whether the entity that is made through SCNT is human life is very controversial. Second, the protection of the human body may be grounds for restriction of human embryonic stem cell research. However, if human oocytes or somatic cells themselves are not considered to be human beings, this research doesn't harm human bodies, except the donors'. The problems that are raised by donation can be solved through a valid informed consent process.[86] When we consider the restriction of human embryonic stem cell research, we should balance the related rights, and the party who argues to restrict the research should prove the reasonable legal grounds for restriction. In conclusion, I think there is no strong reason for banning human embryonic stem cell research.

---

82 It is said that there was no serious policy attention paid to research with living human subjects before the Nuremburg trials. John A. Robertson, supra note, fn. 130.

83 Informed consent of the subject is such a restriction.

84 Korean Supreme Court, 1967. 12. 26, 67다591.

85 Human cloning is almost always discussed with the issue of human dignity. For details, see The Report of the President's Council on Bioethics, Human Cloning and Human Dignity, Public Affairs: New York, 2002.

86 For more discussion about the issues of constitutional restriction of human embryonic stem cell research, see Kyu Won Jung, A Legal Discussion on Human Embryonic Cell Cloning by Somatic Cell Nuclear Transfer Technique, Journal of the Korean Bioethics Association Vol. 1 No. 1, 21.

Human embryonic stem cell research using surgically removed ovaries can be performed under some conditions. However, there is one thing to discuss. One of the main factors which may make human embryonic stem cell research successful is the quality of oocytes. From this point of view, surgically removed ovaries have disadvantages in the making of human stem cells, because they have immature or poor oocytes. However, using surgically removed ovaries can allow the researchers to perform human embryonic stem cell research and increase the chances of developing research techniques and even the potential to succeed. In considering and balancing all these factors it is ethically and legally justifiable to use surgically removed ovaries for human embryonic stem cell research.

In summary, human embryonic stem cell research raises many ethical and legal issues and some people try to ban the research. However, it has many merits and there is no strong evidence requiring the ban of this research. No research should be banned, if there is no obvious ethical or legal evidence to justify the prohibition.

The shortage of oocytes is an important problem in performing human embryonic stem cell research and sometimes it causes the problem of harm to women's bodies. Using oophorectomy specimens for human embryonic stem cell research may be explored because it is less risky and less ethically troubling source for human oocytes. Of course, although using surgically removed ovaries is not the first choice for human embryonic stem cell research because of low scientific efficiency, it may be helpful for developing techniques and even increasing the success rate of the research.

B. Informed Consent Process

Informed consent is a core requirement in biomedical research that uses human subjects. "The voluntary consent of the human subject is absolutely essential."[87] Human embryonic stem cell research is no exception. Informed consent in biomedical research is based on the autonomy of subjects.[88] That means informed consent is required mainly for protecting the autonomy of the human subject. Obtaining appropriate informed consent is required by the U.

87 Nuremburg Code.

88 Daniel Strous, Informed Consent to Genetic Research on Banked Human Tissue, 45 Jurimetrics J. 135(2005) at 138; R. Korobkin, Autonomy and Informed Consent in Nontherapeutic Biomedical Research, 54 UCLA L. Rev. 605, at 606.

S. FDA[89] as well as the Bioethics and Biosafety Act of Korea.[90] Informed consent is designed to provide human subjects with all the related information that is needed when they try to decide autonomously whether to participate in a certain experiment.[91] When researchers try to obtain the informed consent of a subject, they should explain appropriate information about the research, such as the nature and purpose of that research, the procedures, the benefits and risks, etc.[92]

Informed consent is one of the most important issues in human embryonic stem cell research. In human embryonic stem cell research using SCNT, there are two types of donors: somatic cell donors and oocyte donors.

Many authors have focused on oocyte donation when they have dealt with informed consent in human embryonic stem cell research. There are some reasons why oocyte donation attracts attention. First, oocyte is a type of reproductive cell. Many think that manipulation or modification of somatic cells may be allowed under certain conditions, but manipulation or modification of gametes must be regulated more strictly. Second, oocyte donation may harm women's bodies. Oocyte donation is an invasive procedure and sometimes it requires the stimulation of ovary by gonadotropins that may cause side effects such as ovarian hyperstimulation syndrome. However, the consent process for somatic cell donation should also be scrutinized carefully. When human embryonic stem cell research using SCNT is performed, the genetic materials of a donated somatic cell will remain after the fusion, while most of the genetic materials of a donated oocyte will be removed.

Using ovaries obtained from patients undergoing surgery for treatment has some advantages compared to using oocytes retrieved for the sole purpose of research, because *in vivo* extra stimulation or maturation of ovaries will not be needed. Considering these points, the informed consent process for donation of oophorectomy specimens might be different from that of oocytes purely for research use.[93]

Informed consent by the patients for using surgically removed ovarian tissue for human embryonic stem cell research is a must. Adequate information and the lack of coercion are essential for valid informed consent.

89 R. Korobkin, supra note 5, at 610.

90 Bioethics and Biosafety Act §5.

91 45 C.F.R. §46.116.

92 For details, see 45 C.F.R. §46.116(a) and Bioethics and Biosafety Act §15.

93 The informed consent process for using surgically removed ovaries is similar to that of using residual oocytes that are obtained during the assisted fertilization in some points. But it is different from using residual oocytes because there is no stimulation or maturation of the ovaries.

However, there is still the question of whether the patient is really free to make her own decision if informed consent for the donation of resected ovaries that are removed during the surgery is obtained by the surgeon himself. For example, the patient may think that refusing the donation may negatively influence her doctor causing her not to get a good treatment. In most cases, this is not a situation of direct coercion but can in effect influence the patient's decision-making.[94] Therefore a disconnection between the oophorectomy performed by a surgeon and the informed consent process for donating the oophorectomy specimen is needed.

The first step of informed consent for using surgically removed ovaries in human embryonic stem cell research is to obtain informed consent for surgery. The doctor in charge should provide the information about the treatment and obtain informed consent from the patient about the oophorectomy. He should also ask the patient whether she will participate in the human embryonic stem cell research. If the patient agrees to participate in that research, the doctor should provide relevant information about that research which is enough to make the informed consent valid and it must be clear that refusal to participate in that research will not affect the surgery. The doctor should provide all the relevant information about that research as usual for informed consent cases, such as the nature and the purpose of the research, advantages and disadvantages of the research, expected side effects of the donation, etc., but he doesn't need to explain the information that is related to ovarian stimulation and the invasive procedure to obtain oocytes because the ovary will be removed during the oophorectomy for treatment.

The second step of the informed consent process is for the researchers who are intending to use the surgically removed ovary for human embryonic stem cell research to obtain the informed consent of the donor. The researchers should provide all the relevant information again to the patients who are willing to donate the surgically removed ovary. When the information is provided it must include the fact that the patient's genetic information may be disclosed, that means if the stem cell line is established it may contain the genetic information of the donor forever, the patient and her relatives will not directly benefit from that research and treatment of diseases using human stem

94 It is not easy to draw a line between the coercion that makes informed consent invalid and the influence that does not corrupt the validity of informed consent. For details, see Ruth R. Faden and Tom L. Beauchamp, A History and Theory of Informed Consent, New York: Oxford University Press, 1986, Chapter. 10; Kyu Won Jung, Informed Consent of the Subject – Voluntariness - , Hanyang Law Review Vol. 23 No. 1, 181(2006).

cells is a long way off. Also, potential benefits of that research should be provided based on the current scientific knowledge.[95]

Besides the two steps of the informed consent process, an IRB member or other independent party should review the informed consent process. In this step, the important point is whether the patient's decision was made freely and based on adequate information. The person in charge of this step should be independent from the surgeon in charge of the patient's treatment as well as the researchers who want to use the surgically removed ovary. The reviewer also should look into how the patient became aware of human embryonic stem cell research and how she got the relevant information. This step is important for two reasons. First, the surgeon in charge and the researchers using the removed ovary may be connected and have a shared interest. Second, ordinary people may have inaccurate knowledge and unrealistic expectations about human embryonic stem cell research. By reviewing previous informed consent processes, we can enhance the transparency of the process and make sure the consent was made freely and based on adequate information.

Even through all these processes, it is very hard to obtain the ideal form of informed consent due to the patient's difficulties in understanding scientific knowledge,[96] the uncertainty of current knowledge of human embryonic stem cell research, miscommunication between patient-subject and doctors or researchers, etc. However, even though there may be doubt, the informed consent process described is still meaningful. Through these processes the patient-subject will recognize her decision is respected and what the using of removed ovary for human embryonic stem cell research means. On the researchers side this process will ensure time to consider the donor's motivation and can also at least have the obvious preventative function of the unjustifiable use of surgically removed ovaries for human embryonic stem cell research.

In summary, the main concern about using surgically removed ovaries for human embryonic stem cell research is how to get valid informed consent. We propose the multistep process that ensures the autonomy of the patient-subject as follows. First, the surgeon in charge of performing the surgical procedure should obtain informed consent from the patient. Second, the researchers who

95 It is debatable whether the surgeon who operates the oophorectomy can do human embryonic stem cell research using the ovaries that he removed. According to the Bioethics and Biosafety Act in Korea, the government appoints the institutes that can receive oocytes and the institutes that want to do human embryonic research should register separately. See Bioethics and Biosafety Act §14, §18.

96 For details, see Kyu Won Jung, Informed Consent of the Subject – Understanding - , Hanyang Law Review Vol. 22 No. 2, 209(2005).

want to use the surgically removed ovaries for their research should independently get informed consent from the same patient. Finally, both consenting processes should be scrutinized by an independent reviewer.

III. CONCLUSION

Using surgically removed ovarian tissue for human embryonic stem cell research should be explored because they have great promise for providing less risky and less ethically troubling sources for human oocytes. It is technologically feasible to employ the retrieval and *in vitro* maturation of immature human oocytes taken from surgically removed ovaries for somatic cell nuclear transfer. Moreover, the proposed consenting process would further ensure the autonomy of the patient-subject. Finally, researchers should be allowed to explore all avenues to search cures. They are entitled to the freedom of expression as much as others.

Human stem cell research is expected to be a great treatment method although there are still many scientific, ethical and legal issues remaining to be solved. Also, there are many kinds of human stem cell research and new methods are being designed every day. We should not forego any type of research, which can be pursued under adequate oversight, until we get to know which one is the best one. We will get more knowledge only after we do the research and the information we get through that research may help to develop effective methods and give us new insight.

INDEX

A

B

C

F

G

H

I

J

K

L

M

N

O

P

Q

R

S

T